Applications, Basics, and Computing
of
Exploratory Data Analysis

Applications, Basics, and Computing of Exploratory Data Analysis

Paul F. Velleman

Cornell University

David C. Hoaglin

Abt Associates Inc. and Harvard University

Duxbury Press

Boston, Massachusetts

PWS PUBLISHERS

Prindle, Weber & Schmidt · ☙ · Willard Grant Press · **wg** · Duxbury Press · ♠
Statler Office Building · 20 Providence Street · Boston, Massachusetts 02116

PWS Publishers is a division of Wadsworth, Inc.

Printed in the United States of America
4 5 6 7 8 9—85 84 83

*The writing of this work, together with some of the
research reported herein, was supported in part by the
United States Government.*

Library of Congress Cataloging in Publication Data

Velleman, Paul F   1949–
    Applications, basics, and computing of explora-
tory data analysis.
    Includes index.
    1. Statistics—Data processing.   2. Mathematical
statistics—Data processing.   I. Hoaglin, D. C.,
joint author.   II. Title.
QA276.4.V44   001.4'225'02854   80-16193
ISBN 0-87150-409-X
(previously ISBN 0-87872-273-4)

ISBN 0-87150-409-X

(previously ISBN 0-87872-273-4)

To

John W. Tukey

Contents

The BASIC programs in this book are available in machine-readable form from CONDUIT, P.O. Box 388, Iowa City, Iowa 52244 (319)353-5789.

The FORTRAN programs in this book are available in machine-readable form from CONDUIT and from International Mathematical & Statistical Libraries, Inc., 6th Floor, NBC Building, 7500 Bellaire Boulevard, Houston, Texas 77036 (713)772-1927.

A version of the BASIC programs tailored for the Apple micro computer is available from CONDUIT.

Preface

Exploratory data analysis techniques have added a new dimension to the way that people approach data. Over the past ten years, we have continually been impressed by how easily they have enabled us, our colleagues, and our students to uncover features concealed among masses of numbers. Unfortunately, the diversity of these techniques has at times discouraged students and data analysts who may want to learn a few methods without studying the full collection of exploratory tools. In addition, the lack of precisely specified algorithms has meant that computer programs for these techniques have not been widely available. This software gap has delayed the spread of exploratory methods.

We have selected nine exploratory techniques that we have found most often useful. Each of these forms the basis for a chapter, in which we

- Lay the foundations for understanding the technique,
- Describe useful variations,
- Illustrate applications to real data, and
- Provide computer programs in FORTRAN and BASIC.

The choice of languages makes it very likely that at least one of the programs for each technique can be readily installed on whatever computer system is available, from personal microcomputers to the largest mainframe.

Most of this book requires no college-level mathematics and no more than an introduction to statistical concepts. It can serve as a supplementary text to introduce the ideas and techniques of exploratory data analysis into a beginning course in statistics. (In draft form we have used portions of the book in just this way.) Some chapters include advanced sections which assume some knowledge of statistics and are intended to relate the exploratory techniques to traditional statistical practice. These sections will be of greater interest to researchers who wish to use the methods and programs in their own data analysis. A reader who is primarily interested in computational aspects of exploratory data analysis will find both the essential details and many refinements in our programs. At the other extreme, a student who has no background in programming and no access to a computer should have no difficulty in learning the techniques and applying them by pencil and paper. Between these two extremes, the reader who has access to the Minitab statistical system can take immediate advantage of our programs because they have been incorporated into Minitab (Releases 81.1 and later).

Acknowledgments

We are deeply grateful to the colleagues and friends who encouraged and aided us while we were developing this book. John Tukey originally suggested that we provide computer software for exploratory data analysis; later he participated in formulating the new resistant-line algorithm in Chapter 5, and he gave us critical comments on the manuscript. Frederick Mosteller gave us steadfast encouragement and invaluable advice, helped us to aim our writing at a high standard, and made many of the arrangements that facilitated our collaboration. Cleo Youtz painstakingly worked through the manuscript and helped us to eliminate a number of errors, large and small. John Emerson, Kathy Godfrey, Colin Goodall, Arthur Klein, J. David Velleman, Stanley Wasserman, and Agelia Ypelaar read various drafts and contributed helpful suggestions. Stephen Peters, Barbara Ryan, Thomas Ryan, and Michael Stoto gave us critical comments on the programs. Jeffrey Birch, Lambert Koopmans, Douglas Lea, Thomas Louis, and Thomas Ryan reviewed the manuscript and suggested improvements. Teresa Redmond typed the manuscript, and Evelyn Maybee and Marjorie Olson typed some earlier draft material.

We also appreciate the support provided by the National Science Foundation through grant SOC75–15702 to Harvard University.

Initial versions of some BASIC programs were developed on a Model 4051 on loan from Tektronix, Inc.

Introduction

One recent thrust in statistics, primarily through the efforts of John Tukey, has produced a wealth of novel and ingenious methods of data analysis. In his 1977 book, *Exploratory Data Analysis,* and elsewhere, Tukey has expounded a practical philosophy of data analysis which minimizes prior assumptions and thus allows the data to guide the choice of appropriate models. Four major ingredients of exploratory data analysis stand out:

- *Displays* visually reveal the behavior of the data and the structure of the analyses;
- *Residuals* focus attention on what remains of the data after some analysis;
- *Re-expressions,* by means of simple mathematical functions such as the logarithm and the square root, help to simplify behavior and clarify analyses; and
- *Resistance* ensures that a few extraordinary data values do not unduly influence the results of an analysis.

This book presents selected basic techniques of exploratory data analysis, illustrates their application to real data, and provides a unified set of computer programs for them.

The student learning exploratory data analysis (EDA) soon becomes familiar with many pencil-and-paper techniques for data display and analysis. But computers have become valuable aids to data analysis, and even in EDA we may want to turn to them when:

- We have already acquired a feel for the working of a method and want to concentrate on the results rather than the arithmetic;
- We face a large amount of data;
- We want to eliminate tedious arithmetic and the errors that inevitably creep in;
- We want to combine exploratory methods with other data analytic techniques already programmed.

This book shows how we can use the computer for exploratory data analysis. Exploratory methods, however, call for frequent application of the analyst's judgment, and this judgment cannot readily be cast in simple rules and plugged into computer programs. In developing the algorithms in this book, we have often had to give precise rules for judgments such as determining which scale makes a display "look nice," finding points "representative" of a part of the data, or terminating an iterative procedure. In choosing these, we have tried to preserve the underlying resistant features of EDA. For example, the precept that an extraordinary data value should not unduly influence an analysis has led to displays whose message cannot be ruined by such points.

At times the beauty of EDA can be marred by the limitations of the computer. Choices other than our rules and heuristics are possible and may be preferable in some situations. We have tried to offer opportunities to overrule the programs' default decisions. We have also presented the pencil-and-paper versions of the techniques to encourage readers to work by hand when possible and to be aware of the constraints of the computer environment otherwise.

After studying the examples and gaining experience with the EDA techniques, readers who already know some statistics may want to learn more about how an EDA technique compares with a similar traditional method. In some chapters, a starred section (indicated by a * at the section heading) provides brief background information. Generally, a full comparative discussion would involve statistical theory.[1]

The variety of approaches, as well as the alternative analyses that we present for some sets of data, serves to emphasize that practical applications of data analysis generally do not lead to a single "correct" answer. The analyst's judgment and the circumstances surrounding the data also play important roles.

Each chapter also contains a short discussion of programming details (indicated by a † at the section heading), including the algorithm used by the program, alternative methods, and potential implementation difficulties. This section of the chapter, intended primarily for readers interested in statistical

---

[1]Such discussions are the subject of *The Statistician's Guide to Exploratory Data Analysis*, now being prepared under the editorship of David Hoaglin, Frederick Mosteller, and John Tukey.

computing and for instructors, provides necessary background and aids in installing the programs.

Readers of the programs and background discussions should have some knowledge of computing, an acquaintance with EDA and, for some sections, a knowledge of statistics. Readers intending to install the programs are advised to follow a different path, or thread, through the book, and read chapters not in the order natural for learning exploratory data analysis but in the order easiest for understanding the programs.

This book, then, has two main audiences, and each will thread its way through the chapters in a quite different order; so we think of this book as a *threaded text*. Students of exploratory data analysis, researchers intending to use EDA methods, and especially readers who already have the programs available to them on a computer can use the thread that follows the chapters in order, skip the (†) sections of program listings and technical discussions, and select the statistically advanced (*) sections that suit them. For programmers, the thread is best described by the following order of chapters:

C   Programming Conventions
B   Utility Programs
2   Letter-Value Displays
7   Coded Tables
A   Computer Graphics
3   Boxplots
1   Stem-and-Leaf Displays
4   *x-y* Plotting (condensed plots)
5   Resistant Line
6   Smoothing Data
8   Median Polish
9   Rootograms
D   Minitab Implementation

Programmers will find toward the end of most chapters a signpost like this

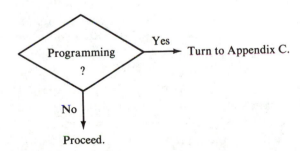

to help them follow the thread. Indeed, they should follow this signpost now.

Note to the Student

If you have not used a computer before, we must warn you that despite our efforts to write simple programs, the programs we give may not run without change on your computing system. Unfortunately, all computing systems are different, and few sophisticated programs can be run on many different systems and remain readable. Therefore, you may need help from an expert on your particular computing system, and he or she will find assistance in the appendices of this book. If the programs already work on your computing system, you will still need to learn the local conventions for using them. This book tells you how to control an analysis procedure, but local conventions will determine how you actually talk to the machine to tell it what to do.

In your first experience with a computer, you must remember that the computer is not doing anything you do not already know how to do by hand (or will know by the time you get to that chapter)—the computer just works more quickly and more accurately. All the same, the machine is stupid, and occasionally you will want to modify its programmed decisions so as to make a display look different or make an analysis work in a different way. Many chapters show you how the modification can be done. We hope that, by relieving you of tedious hand computation and hand graphing, we will free you to interpret the results of the analyses and understand how the methods work.

Note to the Instructor

Many of the chapters in this book can fit in nicely as supplements to an introductory statistics course. In our teaching we have found stem-and-leaf displays and letter-value displays very useful at the start of an introductory course. Boxplots are a useful accompaniment to the comparison of groups.

The resistant line serves as an excellent introduction to simple regression. It provides an elementary yet well-defined method of fitting a line to x-y data, and it offers the pedagogical advantage of a slope formula in the standard form of "change in y divided by change in x." The contrast between resistant lines and least-squares lines helps students to understand the usefulness and limitations of each.

We commonly use boxplots again to introduce one-way analysis of variance. Coded tables and median polish serve as an excellent introduction to the additive structure of two-way analysis of variance. Here, as with regression, we find that teaching the exploratory method first makes the least-squares methods easier to understand.

We have also used EDA to introduce ideas less common in introductory courses. First, we think it is valuable to present more than one method for important statistical models. This counteracts the impression that there is one and only one correct way to analyze data, and it promotes understanding of the strengths and weaknesses of different methods. We have consistently found it valuable to teach data re-expression even in the most elementary courses, and we encourage instructors to use those parts of Chapters 2, 5, and 8. We have also found that the identification and discussion of outliers (Section 3.3) is a useful part of an introductory course.

Exhibits 1 and 2 present two outlines for merging EDA methods with traditional introductory material. The first follows a traditional sequence, while the second follows a topic sequence that puts less emphasis on probability theory and more on data analysis.

The programs themselves are given in two programming languages, FORTRAN and BASIC. While many students will not study the programs in detail, they may find them handy for reference, and we have taken great care to make them as readable and portable as language restrictions permit. As we explain further in Appendix C, the FORTRAN programs satisfy the standards of the PFORT Verifier, which embodies a restricted and almost universally portable subset of the FORTRAN language. They also generally conform to the algorithm standards of *ACM Transactions on Mathematical Software* and *Applied Statistics*. The BASIC programs have been designed for maximum portability to small computers (although BASIC has no standard language definition comparable to PFORT).

Exhibit 1   Outline for Integrating EDA into a Traditional Sequence (EDA topics in italics)

Introductory Comments
    (What is statistics, etc.)
    (Notation)
Describing Distributions of Measurements
    Stem-and-leaf displays
    Histograms
    Measures of central tendency
    Measures of variability
    Letter-value displays
        Re-expressing data to improve symmetry (optional)
Probability
Random Variables and Probability Distributions

Exhibit 1   (continued)

The Binomial Probability Distribution
The Normal Probability Distribution
    The central limit theorem
    Comparing a sample to the normal distribution (Section *2.6, optional)
Large-Sample Statistical Inference
    Point estimation of a population mean
    Interval estimation of a population mean
    Simple boxplots
    Estimating the difference between two means
    Comparing boxplots
        Notched boxplots (optional)
    Hypothesis testing
Inference from Small Samples
    Student's *t*
Linear Regression and Correlation
    Resistant line
    The method of least squares
    Inferences for least-squares regression coefficients
    Re-expressing to straighten a relationship (Section 5.7, optional)
    The correlation coefficient
    Comparing resistant lines and regression lines
Analysis of Enumerative Data
    Tables of data
    Coded tables
    Chi-squared test
The Analysis of Variance
    A comparison of more than two means
    Multiple boxplots
    One-way ANOVA
    Median polish and the additive two-way model
    Two-way ANOVA
Time Series
    Nonlinear data smoothing
    Models for time-series data

Exhibit 2   Outline for Integrating EDA into a "Terminal" Course (EDA topics in italics)

Introductory Comments
    (What is statistics, etc.)
    (Notation)
Describing Distributions of Measurements
    Stem-and-leaf displays
    Measures of central tendency
    Measures of variability
    Letter-value displays
    Re-expressing data to improve symmetry
    Outliers in data (Sections 3.1 through 3.4)
Fitting Lines to x-y Relationships
    Resistant line
    The method of least squares
    Re-expressing to straighten a relationship
    Examining residuals from a linear fit
Elementary Probability
Inferenees for Large Samples
    Interval estimation for the population mean
    Hypothesis testing
    Estimating the difference between two means
Inference for Small Samples
    Student's *t*
Inferences for Linear Regression
    t-tests for regression coefficients
    Correlation
    Comparing resistant lines and least-squares regression
Analyzing Tables of Data
    Coded tables
    The chi-squared statistic
Additive Models for Tables of Data
    Comparing more than two means
    Multiple (notched) boxplots
    One-way ANOVA
    Median polish
    Two-way ANOVA
Time Series
    Nonlinear data smoothing
    Models for time-series data

Chapter 1

Stem-and-Leaf Displays

batch

Data can come in many forms. The simplest form is a collection, or *batch,* of data values. While we probably know something about the data, we are usually wise to assume little at first and just examine the data. Exploratory data analysis provides tools and guidelines for getting acquainted with the data.

display

The first step in any examination of data is drawing an appropriate picture or *display*. Displays can show overall patterns or trends. They also can reveal surprising, unexpected, or amusing features of the data that might otherwise go unnoticed.

stem-and-leaf

The *stem-and-leaf* display has all of these virtues and can be constructed and read easily. With it we can readily see:

- How wide a range of values the data cover;
- Where the values are concentrated;
- How nearly symmetric the batch is;
- Whether there are gaps where no values were observed;
- Whether any values stray markedly from the rest.

These are features that might go unnoticed if we looked no deeper than the data values.

1

In a stem-and-leaf display, the data values are sorted into numerical order and brought together graphically. When we work by hand, we can combine these operations into a single process. When the data have been entered into a computer, a stem-and-leaf display brings the individual values back into view in a way that helps us to see important patterns.

1.1   Stems and Leaves

The basic idea of a stem-and-leaf display is to let the digits of the data values themselves do most of the work of sorting the batch into numerical order and displaying it. A certain number of the digits at the beginning of each data value serve as the basis for sorting, and the next digit appears in the display. According to rules to be explained shortly, we split each data value into its leading digits and its trailing digits. For example, the rules might tell us to split 44,360 as shown in the sketch.

leading digits	trailing digits
44	360
use in sorting	show in display

The leading digits of 44,360 would then be 44, and the trailing digits would be 360. The leftmost trailing digit, 3, would appear in the display to represent this data value. By treating a whole batch of data in this way, we form a stem-and-leaf display.

Before turning to the procedure for constructing a stem-and-leaf display, let us look at the overall appearance of a simple example. Exhibit 1–2 illustrates a simple stem-and-leaf display for the data in Exhibit 1–1. The leading digits appear to the left of the vertical line, but are not repeated for each data value. The leftmost trailing digit of each data value appears to the right of the vertical line.

We construct a stem-and-leaf display in the following steps:

Exhibit 1–1   Acid Levels in Precipitation

Date of Event	pH
20 Dec. 1973	4.57
25–26 Dec. 1973	5.62
30 Dec. 1973–1 Jan. 1974	4.12
9 Jan. 1974	5.29
18–19 Jan. 1974	4.64
21 Jan. 1974	4.31
26–27 Jan. 1974	4.30
28 Jan. 1974	4.39
6–7 Feb. 1974	4.45
9–11 Feb. 1974	5.67
16–17 Feb. 1974	4.39
23–24 Feb. 1974	4.52
24–25 Feb. 1974	4.26
28 Feb. 1974–1 Mar. 1974	4.26
8 Mar. 1974	4.40
9 Mar. 1974	5.78
15–16 Mar. 1974	4.73
21 Mar. 1974	4.56
29–31 Mar. 1974	5.08
3–4 Apr. 1974	4.41
7–9 Apr. 1974	4.12
14 Apr. 1974	5.51
25–26 Apr. 1974	4.82
11–12 May 1974	4.63
17 May 1974	4.29
23 May 1974	4.60

Source: Reported by J.O. Frohliger and R. Kane, "Precipitation: Its Acidic Nature," *Science* 189 (8 August 1975):455–457 from samples collected at a location in Allegheny County, Pennsylvania. Copyright 1975 by the American Association for the Advancement of Science. Reprinted by permission.

Note: pH is an alkalinity/acidity measure. A pH of 7 is neutral; values below 7 are acidic.

1. Choose a suitable pair of adjacent digit positions in the data and split each data value between these two positions. In going from Exhibit 1–1 to Exhibit 1–2, we have split data values so that the first two digits of each value are the leading digits.
2. Write down a column of all the possible sets of leading digits in order from lowest to highest. These are the stems. (Note that we must include sets of

Exhibit 1–2   Stem-and-Leaf Display for the Precipitation pH Data of Exhibit 1–1

```
                              41 | 22
                              42 | 669
                              43 | 1099
                              44 | 501
                              45 | 726
                              46 | 430
                              47 | 3
                  Stems       48 | 2         Leaves
                              49 |
                              50 | 8
                              51 |
                              52 | 9
                              53 |
                              54 |
                              55 | 1
                              56 | 27
                              57 | 8
```

leading digits that *might* have occurred, but don't happen to be present in this particular batch. Of course, we needn't go beyond the lowest and highest data values.)

3.   For each data value, write down the first trailing digit on the line labeled by its leading digits. These are the leaves, one leaf for each data value.

Let us now see how these steps produce the display in Exhibit 1–2 from the data in Exhibit 1–1.

The data in Exhibit 1–1 report the acidity of 26 samples of precipitation collected at a location in Allegheny County, Pennsylvania, from December 1973 to June 1974. The data are pH values—pH 7 is neutral; lower values are more acidic. They could bear on the theory that air pollution causes rainfall to be more acidic than it would naturally be.

Exhibit 1–2 shows the stem-and-leaf display of these values. To make the display, we must split each number into a stem portion and a leaf portion. For the stem-and-leaf display in Exhibit 1–2, the pH values were split between the tenths digit and the hundredths digit. For example, the entry in Exhibit 1–1 for 20 Dec. 1973, which is 4.57, became 45|7, so that the stem is 45 and the

Exhibit 1–3 Full Stem-and-Leaf Display for the Precipitation pH Data of Exhibit 1–1

Unit = .01
1 2 represents 0.12

2	41	22
5	42	669
9	43	1099
12	44	501
(3)	45	726
11	46	430
8	47	3
7	48	2
	49	
6	50	8
	51	
5	52	9
	53	
	54	
4	55	1
3	56	27
1	57	8

leaf is 7. Working from the data in Exhibit 1–1 and writing down the leaves as we read through the data in order yield the display in Exhibit 1–2. In the second line, we can easily verify that 42|669 stands for the three data values 4.26, 4.26, and 4.29.

Choosing the pair of adjacent digit positions for the stem-leaf split is basically a matter of straightforward judgment, and easily learned. However, because the location of the decimal point is lost when we split the data values into stems and leaves, the finished version of the display should include a reminder of where the decimal point falls. This reminder is usually provided in *unit* a heading above the display by declaring the *unit* as the decimal place of the leaf, and by providing an example.

Exhibit 1–3 shows a more elaborate version of the basic stem-and-leaf display of Exhibit 1–2. This version is the standard form of the stem-and-leaf display. Here the heading specifies the unit (.01) and gives an example, "1 2 represents 0.12," so that we can tell that 42|669 represents 4.26, 4.26, and 4.29, rather than, say, 42.6, 42.6, and 42.9.

depths Exhibit 1–3 also includes a column of *depths* located to the left of the stem column. In the depth column, the number on a line tells how many leaves lie either on that line or on a line closer to the nearer end of the batch. Thus, the 5 on the second line of Exhibit 1–3 says that five data values fall either on that line or closer to the low-pH end of the batch; actually, three values—4.26, 4.26, and 4.29—are on the second line, and two—4.12 and 4.12—are on the first line. Naturally, the depths increase from each end toward the middle of the batch.

The depth information is shown differently at the middle of the batch. The line containing the middle value shows a count of *its* leaves in the depth column, enclosed in parentheses. When the batch has an even number of data values, no single value will be exactly in the middle. Instead, a pair of data values will surround the middle. If this happens, and each middle value falls on a different line, the depths are shown as usual. Chapter 2 discusses depths and shows how they help in finding values to summarize the data.

Exhibit 1–3 reveals several features of the precipitation pH data: Most of the values form a broad group from 4.1 to 4.7; scattered values trail off above that group to 5.29; and four values form a clump from 5.51 to 5.78. On these four occasions the precipitation was noticeably less acidic than at other times—a feature we would not have seen without a display.

As we have seen in Exhibit 1–3, a stem-and-leaf display helps to highlight a variety of features in a batch of data. When we need to identify individual data values, we can do so because the numbers themselves form the display. This can make it easier for the data analyst to decide which features are important and what they mean in the context of the data.

1.2 Multiple Lines per Stem

To produce an effective display for any batch we encounter, we must have ways of stretching out a display that looks squeezed onto too few lines and of squeezing together a display that looks stretched out over too many lines. We can improve the appearance of a stem-and-leaf display by splitting stems into either two equal parts or five equal parts and by using one line for each part.

In the simplest type of stem-and-leaf display, such as Exhibit 1–3, all ten digits, 0 through 9, can be used as leaves on each line. When stretching out a display to use two lines per stem, we place leaf digits 0, 1, 2, 3, and 4 on the first line (indicated by a * after the stem) and 5, 6, 7, 8, and 9 on the second line (indicated by a ·), and thus produce a variation of the original simple display using twice as many lines. Exhibit 1–5 shows an example of 2-line stems based on the data in Exhibit 1–4. The numbers in this display are the relative air pollution potentials of hydrocarbons (HC) in 60 U.S. cities (actually Standard Metropolitan Statistical Areas, SMSAs). For example, the first line in Exhibit 1–5 represents the hydrocarbon pollution potentials for Dallas, Fort Worth, Miami, New Haven, and Wichita. This display illustrates an additional useful variation: listing apparently stray values on a separate line, labeled "HI" for high strays. Section 1.4 discusses this variation further.

When we use five lines per stem, we find that it helps—both in making a stem-and-leaf display by hand and in reading one already made—to have a distinctive label on each line. We place leaves 0 and 1 on a line labeled *, leaves 2 and 3 on the T (for Two and Three) line, leaves 4 and 5 on the F (Four and Five) line, leaves 6 and 7 on the S line, and leaves 8 and 9 on the · line. We can think of this display as using five times as many lines as the simple display. More commonly, however, the 5-line display is a way of using *half* as many lines: We first move the split between stem and leaf one digit position to the left and then use five lines per stem. Exhibit 1–6 shows the precipitation pH data in this way. The split between stem and leaf has been shifted left to the decimal point so that the final digit of each value is omitted and the second digit serves as the leaf. For example, the first line in Exhibit 1–6 represents the same data values as the first line in Exhibit 1–3—that is, pH 4.12. In Exhibit 1–6 the tenths digit is the leaf; in Exhibit 1–3 the tenths digit is part of the stem. The hundredths digit, 2, is not used in Exhibit 1–6. The shape of the main body of numbers (lines 4* through 4S) is now easier to see, but the 4 less acidic precipitation samples are not as prominent. Our choice of scale in stem-and-leaf displays usually depends on what kinds of patterns are most important to us as we examine the data.

When, as in Exhibit 1–6, the unit in the stem-and-leaf display is not

Exhibit 1–4   Four Variables for 60 U.S. SMSAs

SMSA	January Mean Temperature °C	HC Pollution Potential	Median Education	Age-Adjusted Mortality
Akron, OH	−2.78	21	11.4	921.87
Albany, NY	−5.00	8	11.0	997.87
Allentown, PA	−1.67	6	9.8	962.35
Atlanta, GA	7.22	18	11.1	982.29
Baltimore, MD	1.67	43	9.6	1071.29
Birmingham, AL	7.22	30	10.2	1030.38
Boston, MA	−1.11	21	12.1	934.70
Bridgeport, CT	−1.11	6	10.6	899.53
Buffalo, NY	−4.44	18	10.5	1001.90
Canton, OH	−2.78	12	10.7	912.35
Chattanooga, TN	5.56	18	9.6	1017.61
Chicago, IL	−3.33	88	10.9	1024.89
Cincinnati, OH	1.11	26	10.2	970.47
Cleveland, OH	−2.22	31	11.1	985.95
Columbus, OH	−0.56	23	11.9	958.84
Dallas, TX	7.78	1	11.8	860.10
Dayton, OH	−1.11	6	11.4	936.23
Denver, CO	−1.11	17	12.2	871.77
Detroit, MI	−2.78	52	10.8	959.22
Flint, MI	−4.44	11	10.8	941.18
Fort Worth, TX	7.22	1	11.4	891.71
Grand Rapids, MI	−4.44	5	10.9	871.34
Greensboro, NC	4.44	8	10.4	971.12
Hartford, CT	−2.78	7	11.5	887.47
Houston, TX	12.78	6	11.4	952.53
Indianapolis, IN	−1.67	13	11.4	968.67
Kansas City, MO	−0.56	7	12.0	919.73
Lancaster, PA	0.0	11	9.5	844.05
Los Angeles, CA	11.67	648	12.1	861.83
Louisville, KY	1.67	38	9.9	989.26
Memphis, TN	5.56	15	10.4	1006.49
Miami, FL	19.44	3	11.5	861.44
Milwaukee, WI	−6.67	33	11.1	929.15
Minneapolis, MN	−11.11	20	12.1	857.62
Nashville, TN	4.44	17	10.1	961.01
New Haven, CT	−1.11	4	11.3	923.23
New Orleans, LA	12.22	20	9.7	1113.16

Exhibit 1–4   (continued)

SMSA	January Mean Temperature °C	HC Pollution Potential	Median Education	Age-Adjusted Mortality
New York, NY	0.56	41	10.7	994.65
Philadelphia, PA	0.0	29	10.5	1015.02
Pittsburgh, PA	−1.67	45	10.6	991.29
Portland, OR	3.33	56	12.0	893.99
Providence, RI	−1.67	6	10.1	938.50
Reading, PA	0.56	11	9.6	946.19
Richmond, VA	3.89	12	11.0	1025.50
Rochester, NY	−3.89	7	11.1	874.28
St. Louis, MO	0.0	31	9.7	953.56
San Diego, CA	12.78	144	12.1	839.71
San Francisco, CA	8.89	311	12.2	911.70
San Jose, CA	9.44	105	12.2	790.73
Seattle, WA	4.44	20	12.2	899.26
Springfield, MA	−2.22	5	11.1	904.16
Syracuse, NY	−4.44	8	11.4	950.67
Toledo, OH	−3.33	11	10.7	972.46
Utica, NY	−5.00	5	10.3	912.20
Washington, DC	2.78	65	12.3	967.80
Wichita, KS	0.0	4	12.1	823.76
Wilmington, DE	0.56	14	11.3	1003.50
Worcester, MA	−4.44	7	11.1	895.70
York, PA	0.56	8	9.0	911.82
Youngstown, OH	−2.22	14	10.7	954.44

Source: G.C. McDonald and J.A. Ayers, "Some Applications of the 'Chernoff Faces': A Technique for Graphically Representing Multivariate Data," in Peter C.C. Wang, ed., *Graphical Representation of Multivariate Data* (New York: Academic Press, 1978), pp. 183–197. Copyright © 1978 by Academic Press, Inc. All right of reproduction in any form reserved. Reprinted by permission.

Note: The data in this exhibit are used in Exhibit 1–5 and in later exhibits.

the last digit position provided in the data, the digits following the unit position do not appear in the display. Even then, individual data items can still be matched easily with leaves because the stems and leaves are the leftmost digits of the numbers. To ensure this, we do not round values when digits are left off, but rather we truncate the data values. That is, we drop trailing digits to preserve the original digits on either side of the stem-leaf split.

Exhibit 1–5   Relative Air Pollution Potential of Hydrocarbons in 60 U.S. SMSAs

Unit = 1
1   2 represents 12.

```
  5      0* 11344
 21      0· 5556666677778888
 30      1* 111122344
 30      1· 577888
 24      2* 000113
 18      2· 69
 16      3* 0113
 12      3· 8
 11      4* 13
  9      4· 5
  8      5* 2
  7      5· 6
         6*
  6      6· 5
```

HI|88,105,144,311,648

Note: Data from Exhibit 1–4.

Exhibit 1–6   A Stem-and-Leaf Display of the Precipitation pH Data in Exhibit 1–1, Using 5 Lines per Stem

Unit = .1
1   2   represents 1.2

```
  2      4* 11
  9      4T 3333222
 (6)     4F 545454
 11      4S 6766
  7      4· 8
  6      5* 0
  5      5T 2
  4      5F 5
  3      5S 667
```

1.3   Positive and Negative Values

When a batch includes both positive and negative values, the stems near zero take a special form. Numbers slightly greater than zero appear on a stem labeled +0. Numbers slightly less than zero appear on a stem labeled −0. This labeling may seem strange at first; we might expect the stem −1 to be next to +0, but a simple example shows why it is necessary. Exhibit 1–7 shows a stem-and-leaf display of the mean January temperatures in degrees Celsius for the 60 U.S. SMSAs in Exhibit 1–4. (Recall that 0°C is the freezing point of water.) In Exhibit 1–7, numbers like −1.1° and −1.6° are placed on the −1

Exhibit 1–7   Stem-and-Leaf Display of Mean January Temperatures in °C at 60 U.S. SMSAs

Unit = .1
1   2   represents 1.2

```
                      LO| − 111

          2          −6|6
          4          −5|00
          9          −4|44444
         12          −3|383
         19          −2|7727722
         28          −1|611116166
         (4)         −0|5500
         28          +0|055055
         22           1|616
         19           2|7
         18           3|38
         16           4|444
         13           5|55
                      6|
         11           7|2272
          7           8|8
          6           9|4

                      HI|127,116,194,122,127
```

Note: Data from Exhibit 1–4.

stem. The -0 stem is needed for numbers like $-0.5°$. The special value 0.0 could be placed on either of the two 0 stems. To preserve the outline of the display, we split the 0.0 values equally between the $+0$ stem and the -0 stem.

In Exhibit 1–7, the major feature is the 41 cities that have mean January temperatures between $-6.6°C$ and $+2.7°C$. One clump of cities— generally those in the Southwest—stands out from $7.2°C$ to $9.4°C$. Five cities—Houston, Los Angeles, Miami, New Orleans, and San Diego—appear on the HI stem; and Miami, at $19.4°C$, is the highest. Minneapolis, at $-11.1°C$, appears on the LO stem.

1.4   Listing Apparent Strays

Data values that stray noticeably from the rest of the batch are a common enough occurrence for us to give them special treatment in stem-and-leaf

---

Exhibit 1–8   Stem-and-Leaf Display of the Hydrocarbon Pollution Potentials in Exhibit 1–5 without the Use of a HI Stem

---

```
Unit = 10
1   2   represents 120.


(52)    0*|0000000000000000000000011111111111111112222222233333444
   8    0·|5568
   4    1*|04
        1·|
        2*|
        2·|
   2    3*|1
        3·|
        4*|
        4·|
        5*|
        5·|
   1    6*|4
```

displays. We want to avoid a display in which most data values are squeezed onto a few lines of the display, the strays occupy a line or two at one or both extremes, and many lines lie blank in between. For example, Exhibit 1–8 shows what the display in Exhibit 1–5 would have looked like if we had not isolated the stray high values.

Once we have decided which data values to treat as strays, we can easily list them separately at the low or high end of the display where they belong. We introduce these lists with the labels LO and HI in the stem column, and we leave at least one blank line between each list and the body of the display in order to emphasize the separation.

When we produce the display by hand, we can usually use our judgment in differentiating strays from the rest of the data. A computer program, however, must rely on a rule of thumb to make this decision in hopes of producing reasonable displays for most batches. This rule is discussed in detail in Chapter 3.

1.5 Histograms

histogram

Data batches are often displayed in a *histogram* to exhibit their shape. A histogram is made up of side-by-side bars. Each data value is represented by an equal amount of area in its bar. We can see at a glance whether the batch is generally *symmetric*—that is, approximately the same shape on either side of a line down the center of the histogram—or whether it is *skewed*—that is, stretched out to one side or the other of the center. We can also see whether a histogram rises to a single main hump—a *unimodal* pattern—or exhibits two or more humps—a *bimodal* or *multimodal* pattern, respectively. The parts on either end of a histogram are usually called the *tails*. We can characterize a histogram as showing short, medium, or long tails according to how stretched out they are. Finally, we can spot straggling data values that seem to be detached from the main body of the data.

symmetric
skewed

unimodal
bimodal
multimodal
tails

Unimodal symmetric batches are usually the easiest to deal with. Multiple humps may indicate identifiable subgroups—for example, male and female—that might be more usefully examined separately. (One way to deal with skewness, or asymmetry, is described in Chapter 2; extraordinary data values are discussed more precisely in Chapter 3.)

The stem-and-leaf display resembles a histogram in that both of them display the distribution of the data values in the batch by representing each value with an equal amount of area. In a stem-and-leaf display, each digit occupies the same amount of space. In a histogram, each data value is represented by an equal amount of area in a bar delineated by lines. Occasionally a histogram is made up of printed symbols by using a single character—typically * or X—to represent each value. (This is done by many computer programs.) For large batches, a single * can represent several data values in a histogram in order to preserve a manageable size. Thus a histogram can serve as an "overflow" alternative to a stem-and-leaf display when the batch is large (several hundred values or so). With several hundred leaves we would be less able to concentrate on detail anyway.

When we can look at the detail, however, the stem-and-leaf display can reveal patterns not found in a histogram. Exhibit 1–9 compares a computer-produced histogram with a stem-and-leaf display. The data are the pulse rates of 39 Peruvian Indians. The outlines of the two are not identical because the histogram is based on a different set of intervals, but this is not the interesting feature of these data. What is interesting is that *all* the leaves in the stem-and-leaf display are even digits (0, 2, 4, 6, 8) and that all the data values except one (74) are divisible by 4. Although the pulse rates were reported in beats per minute, they were probably measured by counting beats for 15 seconds and then multiplying by 4. Perhaps, in the exceptional case (74) the observer overshot the 15-second mark, counted pulses for a further 15 seconds, and multiplied by 2. Such wide spacing of values (in this case, by multiples of 4) creates a granularity that could make a difference in some analyses and would certainly have remained hidden in a histogram.

Exhibit 1–9   Histogram and Stem-and-Leaf Display of the Pulse Rates of 39 Peruvian Indians

MIDDLE OF INTERVAL	NUMBER OF OBSERVATIONS		STEM-AND-LEAF DISPLAY UNIT = 1.0 1 2 REPRESENTS 12.
50.	1	*	1 5* 2
55.	1	*	2 · 6
60.	6	******	15 6* 0000004444444
65.	7	*******	19 · 8888
70.	12	************	(9) 7* 222222224
75.	5	*****	11 · 6666
80.	2	**	7 8* 004
85.	1	*	4 · 888
90.	4	****	1 9* 2

Source: Ryan, T. A., B. L. Joiner, and B. F. Ryan. 1976. *The Minitab Student Handbook* (N. Scituate, Mass.: Duxbury Press) p. 277.

A subtler granularity can be seen in the mean January temperatures in Exhibit 1–7. Inspection of this exhibit reveals that no more than two different leaf values occur on any stem and that the actual values are symmetric around the zero stem. For example, stems 3 and − 3 have only leaves of 3 or 8; stems 1 and − 1 have only leaves of 1 or 6. This granularity occurs because the temperatures originally were recorded to the nearest degree in Fahrenheit and then were converted to Celsius. Patterns of this kind are the ones most likely to be overlooked when data are analyzed on a computer. They highlight an important function of the stem-and-leaf display—keeping the individual data values in view.

1.6   Stem-and-Leaf Displays from the Computer

It is easy to construct a stem-and-leaf display by hand. With a little practice one quickly learns to choose the number of lines per stem that neither stretches out the display too far nor cramps it into too few lines.

It is not nearly as easy to write a general computer program to produce stem-and-leaf displays. Computers cannot follow instructions such as "choose a display format so that the display will be neither too stretched out nor too cramped." Instead, we must devise specific rules that the computer will apply in making the necessary decisions. However, once the program is written, it is easy to use because all the essential decisions can be left to the computer. We need only tell the computer what data we wish it to display. How to do this—and, indeed, how you tell *your* computer to do anything—will depend on the way your computer is set up. If you don't already know how to run the programs in this book on your computer, ask for assistance from someone expert in using it.

Computer-produced stem-and-leaf displays look very nearly the same as hand-produced displays. Since computer output terminals type neatly, a blank column can be used effectively in place of the vertical line to separate stems from leaves, and thus keep the display less cluttered. The heading always states the unit and provides an example because the place at which numbers are split into stems and leaves has been chosen automatically. Exhibit 1–10 shows a computer-printed stem-and-leaf display of the precipitation pH data of Exhibit 1–1. The program has selected the same 5-lines-per-stem scale used in the stem-and-leaf display in Exhibit 1–6 and has identified for the HI stem 3 of the 4 values that appeared to be suspect in Exhibits 1–2 and 1–6. We also see that the leaves are now in numerical order on each stem, whereas they had been in chronological order in the earlier displays.

Exhibit 1–10 Computer-Printed Stem-and-Leaf Display of the
Precipitation pH Data of Exhibit 1–1

```
                                      STEM-AND-LEAF DISPLAY
                                      UNIT = 0.1000
                                      1   2    REPRESENTS 1.2

                               2      4*  11
                               9      T   2223333
                              (6)     F   444555
                              11      S   6667
                               7      4·  8
                               6      5*  0
                               5      T   2
                               4      F   5

                                      HI   56,56,57
```

*program
options*
      Many of the programs in this book include *options* that will allow you to tailor a display or computation to the specific needs of your analysis. One such option is to forbid the use of the HI and LO stems and display all of the data values from lowest to highest in the main body of the stem-and-leaf display. While this is desirable in some situations, the result *may* look like Exhibit 1–8. How you specify this option or any option for any of the programs will, of course, depend on the way your computer is set up.

1.7 Algorithms I

      Although the stem-and-leaf display is one of the simplest exploratory data analysis methods, the stem-and-leaf programs in this chapter are very sophisticated and are among the longest programs in the book. Many decisions must be made when a stem-and-leaf display is created. When we work by hand, we make these decisions so easily that they almost go unnoticed. A program, however, must be prepared for every situation it might face in producing a stem-and-leaf display, and it must specify explicitly how each decision should be made in every situation.

      Several of the decision rules used in the programs at the end of this chapter are subtle and were developed only after considerable trial and error. Some depend upon aspects of data analysis discussed in later chapters. If you are planning to study the programs and the algorithm and have not yet followed the "thread" through Appendices A, B, and C and Chapters 2, 7, and 3, *please* stop and read them first. If you are reading the book in chapter order, please skip the rest of this chapter. When you return to this section after reading the other chapters, you will be able to see how the stem-and-leaf algorithm combines ideas introduced in other chapters and adds new ideas special to this technique.

Note: As discussed in the introduction to this book, programmers will find toward the end of some chapters a direction signpost that will help them thread their way through the book. Here is one:

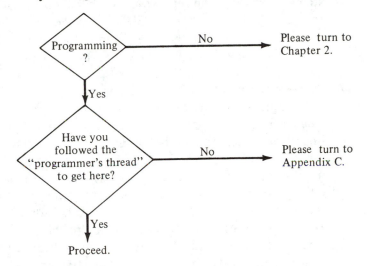

† 1.8   Algorithms II

Stem-and-leaf displays present two problems to the programmer: (1) finding a heuristic algorithm to select the display format and (2) producing a display that is a highly structured combination of numbers, character strings, and numerals based upon numbers. Specifically, each line contains a depth count (treated as a number), a stem (some combination of numbers and characters), and a string of leaves (numerals, with no associated spaces or decimal points, selected from a specific digit position in a number). The programs must be sure to obtain the correct leaf digit (adjusting for the unavoidable rounding error of digital computers). They must keep track of the sign of the data values and of the allocation of data values to lines of the display. Each line is a half-open interval including the inside limit, which is closer to zero and corresponds to a data value whose leaf is zero on that line. The interval extends to, but does not include, the inner limit of the next line away from zero. The zero stems are special because both the $+0$ and -0 stems label intervals that include the value 0.0. The programs must thus pay special attention to zeros in the data.

If the data batch is not already in order, it is first sorted (see Section 2.9 for a discussion of sorting methods). Next, the program must decide whether any extreme data values should appear on the special LO and HI stems. If so, only the remaining numbers will be used in choosing the display format. The details of this decision are discussed in Section 3.3.

The program then determines the unit and the display format by estimating how many lines ought to be used in all to display the numbers. Experience has shown that, if we have n numbers, $10 \times \log_{10} n$ is a good first guess at the number of lines needed for a good display. (Here the number of data values, n, excludes the stray values assigned to the LO and HI stems.) The program first computes the range of values that would be covered by each line if the maximum number of lines were used. This line width is the result of dividing the range of the (non-straying) data values by the approximate number of lines desired ($10 \times \log_{10} n$). Because each line must accommodate either two, five, or ten possible leaf digit values, the line width is rounded up to the next larger number representable as 2, 5, or 10 times an integer power of 10. Rounding up guarantees that no more than $10 \times \log_{10} n$ lines will be used. The power of 10 yields the unit, and the multiplier (2, 5, or 10) is the number of leaf digits on each line. (Note that 10/(number of leaf digits) yields the number of lines per stem.) The program then prints the display heading, which includes the unit decided upon and an example. The example uses a stem of 1 and a leaf of 2 to illustrate where the decimal point should be placed.

Now the program can step through the ordered data and print out one line of the display at a time. The program must print each stem according to the format selected and must use the correct numeral for each leaf. If the leaves to be printed on a line would extend beyond the right margin, the program uses the available spaces and then inserts an asterisk in the rightmost space to show that the overflow occurred. (The depth still provides a complete count and thus indicates the number of values omitted.) These steps require careful programming so that they work for all possible cases.

For each line of the display, the program first looks down the ordered data batch to identify the data values to be displayed on that line. It counts these values and computes the depth, which it places on the output line. It then constructs the stem and places it on the output line. Finally, it scans through the data values and computes and prints leaves. This requires only one pass through the data because one line begins, after allowing for lines that have no leaves, where the previous line ends.

FORTRAN

The FORTRAN programs that produce a stem-and-leaf display consist of five subroutines, STMNLF, SLTITL, OUTLYP, DEPTHP, and STEMP. To produce a stem-and-leaf display for data in the vector Y, use the FORTRAN statement

    CALL STMNLF(Y, N, SORTY, IW, XTREMS, ERR)

where the parameters have the following meanings:

Y( )	is the N-long vector of data values;
N	is the number of data values;
SORTY( )	is an N-long workspace for real numbers;
IW( )	is an N-long workspace for integers;
XTREMS	is a logical flag, set .TRUE. if the plot should include all data values or set .FALSE. to permit HI and LO stems;
ERR	is the error flag, whose values are

0	normal
11	$N \le 1$
12	internal error—see program
13	page has fewer than 5 spaces for leaves.

The subroutine STMNLF first determines the display format. It calls SLTITL to print the headings. If necessary, it then calls OUTLYP to print the LO stem. Then it steps through the sorted data, calling DEPTHP to compute and print depths and STEMP to compute and print stems. STMNLF places the leaves on each line itself. If necessary, it calls OUTLYP to print the HI stem. Throughout, STMNLF uses the utility output routines (see Appendix C).

BASIC

The BASIC subroutine for stem-and-leaf display is entered with the N data values to be displayed in the array Y. If the version number, V1, is 1, the plot is scaled to the extreme values, and no HI and LO stems are printed. If V1 is 2 or greater, extreme values are placed on the HI and LO stems and excluded in determining the plot format. The array Y is returned unmodified. The program uses the defined functions, the SORT subroutines, and the plot-scaling subroutines (see Appendix A).

References

Frohliger, J.O., and R. Kane. 1975. "Precipitation: Its Acidic Nature." *Science* 189 (8 August 1975), pp. 455–457.

McDonald, Gary C., and James A. Ayers. 1978. "Some Applications of the 'Chernoff Faces': A Technique for Graphically Representing Multivariate Data." In Peter C.C. Wang, ed., *Graphical Representation of Multivariate Data*. New York: Academic Press.

Ryan, T.A., B.L. Joiner, and B.F. Ryan. 1976. *The Minitab Student Handbook*. N. Scituate, Mass.: Duxbury Press.

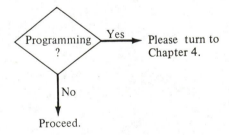

```
5000 REM STEM & LEAF DISPLAY
5010 REM ENTER WITH Y() OF LENGTH N.  M0,M9 ARE LEFT AND RIGHT
5020 REM MARGINS DESIRED.
5030 REM VERSIONS: V1=2 SCALES TO ADJACENT VALUES (NORMAL)
5040 REM           V1=1 SCALES TO EXTREMES
5050 REM CALLS SUBROUTINES(@ LINE):
5060 REM SORT(1000), NPW(1900), YINFO(2500), COPYSORT(3000)
5070 REM
5080 REM SET UP PRINTING DETAILS:  I8 IS POSITION OF STEM/LEAF BREAK

5090 LET I8 = M0 + 11
5100 LET I9 = M9 - I8 - 1
5110 IF I9 > 5 THEN 5140
5120 PRINT "ALLOWED WIDTH OF DISPLAY TOO NARROW"
5130 STOP

5140 REM  SORT Y() TO W() -- (GOSUB 3000 DOES S&L OF X().)

5150 GOSUB 3300

5160 REM  FIND ADJACENT VALUE LOCATIONS FROM PLSCALE

5170 GOSUB 2500

5180 REM  IF ADJACENT VALUES EQUAL, TRY THE EXTREMES

5190 IF A3 = A4 THEN 5210
5200 IF V1 <> 1 THEN 5260

5210 REM SCALE TO EXTREMES--MAY MAKE A BAD DISPLAY.

5220 LET A2 = N
5230 LET A1 = 1
5240 LET A3 = W(1)
5250 LET A4 = W(N)

5260 REM FIND NICE LINE WIDTH

5270 LET A8 = 1
5280 LET P9 = FNI(10 * FNL(A2 - A1 + 1))
5290 LET N5 = 2
5300 LET L0 = A3
5310 LET H1 = A4
5320 GOSUB 1900
```

```
5330 REM NICE WIDTH = N4*10^N3.
5340 REM NOW U= LEAF UNIT. THINK OF ALL VALUES AS INTEGER*10^UNIT.
5350 REM CONVERT TO INTEGERS OF THE FORM S...SL.
5360 REM THE REMAINING WORK CAN BE INTEGER MATH FOR SPEED.
5370 REM KEEP 0 LEAVES ON THE ZERO STEMS CORRECT, SPECIAL TREATMENT
5372 LET W(N + 1) = 0
5374 LET W(N + 2) = 0
5376 LET W(N + 3) = 0
5380 REM FOR NUMBERS SCALED TO 0 COUNT >0,=0,<0 IN W(N+1) TO W(N+3)
5390 LET Z0 = N + 2
5400 FOR I = 1 TO N
5410    LET X1 = FNI(W(I) / U)
5420    IF X1 <> 0 THEN 5440
5430    LET W(Z0 + SGN(W(I))) = W(Z0 + SGN(W(I))) + 1
5440    LET W(I) = X1
5450 NEXT I
5460 LET L0 = W(A1)
5470 LET H1 = W(A2)

5480 REM   SET L9 = LINE WIDTH = NICEWIDTH/UNIT = P7/10^N3 = MANTISSA.

5490 LET L9 = N4
5500 PRINT
5510 PRINT TAB(M0 + 2);"STEM & LEAF DISPLAY"
5520 PRINT TAB(M0 + 2);"   UNIT = ";U
5530 PRINT TAB(M0 + 2);"1  2  REPRESENTS ";
5540 IF U < 1 THEN 5570
5550 PRINT FNI(12 * U)
5560 GO TO 5670
5570 IF U <> .1 THEN 5600
5580 PRINT "1.2"
5590 GO TO 5670
5600 PRINT "0.";

5610 REM   CHECK FOR NON-ANSI BASICS

5620 IF ABS(N3) <= 2 THEN 5660
5630 FOR I = 1 TO ABS(N3) - 2
5640    PRINT "0";
5650 NEXT I
5660 PRINT "12"
5670 PRINT

5680 REM   PRINT VALUES BELOW ADJACENT VALUE. P6=RANK

5690 LET P6 = A1 - 1
5700 IF P6 = 0 THEN 5760
5710 PRINT TAB(I8 - 4);"LO: ";
5720 FOR I = 1 TO P6
5730    PRINT STR$(W(I));", ";
5740 NEXT I
5750 PRINT
5760 PRINT
```

```
5770 REM   INITIALIZE FOR LINE BEFORE THE FIRST LINE.
5780 REM   C0 = LINE CUT. =FIRST NUMBER ON NEXT LINE OF +STEMS,
5790 REM                  =LAST NUMBER ON CURRENT LINE OF -STEMS.
5800 REM   L4 IS STEM PTR = INNER (NEAR ZERO) EDGE OF CURRENT LINE.
5810 REM   N7 IS NEGATIVE FLAG = 1 WHILE STEMS < 0 ,= 0 ELSE.

5820 REM   D0 IS MEDIAN FLAG, = 0 UNTIL MEDIAN IS PAST, =1 AFTER.
5830 REM   K1,K2,K3 ARE POINTERS INTO Y() FOR  DEPTHS, PRINTING, ZEROS.
5840 REM   I1 COUNTS SPACES USED ON THE LINE.
5850 REM   P5 COUNTS LEAVES ON THIS LINE FOR DEPTH CALCULATIONS
5860 REM   L9 IS VALUE COVERED BY ONE LINE
5870 REM   P6 COUNTS RANK, L2 WILL HOLD LEAF DIGIT BELOW

5880 LET C0 = FNF((1 + E0) * L0 / L9) * L9
5890 LET N7 = 1
5900 LET L4 = C0
5910 IF L0 < 0 THEN 5940
5920 LET N7 = 0
5930 LET L4 = C0 - L9
5940 LET D0 = 0
5950 LET K1 = A1
5960 LET K2 = K1

5970 REM PROGRAM CAN BREAK HERE FOR SMALL MACHINES
5980 REM  LOOP: FOR EACH LINE UP TO NUMBER OF LINES

5990 FOR J1 = 1 TO P8

6000    REM  STEP TO NEXT LINE

6010    LET C0 = C0 + L9
6020    IF L4 <> 0 THEN 6080

6030    REM IF THIS WAS THE "-0" STEM,
6040    REM CHANGE THE NEGATIVE FLAG BUT DON'T STEP THE STEM VALUE.

6050    IF N7 = 0 THEN 6080
6060    LET N7 = 0
6070    GO TO 6090
6080    LET L4 = L4 + L9

6090    REM INITIALIZE COUNT OF CHARACTER POSITION ON THE LINE

6100    LET I1 = 0
```

```
6110    REM   FIND AND PRINT DEPTH
6120    REM   NOTE THAT CUT (C0) BEHAVES DIFFERENTLY FOR + AND - STEMS.

6130    LET P5 = 0
6140    FOR K1 = K1 TO A2
6150      IF W(K1) > C0 THEN 6220
6160      IF C0 < 0 THEN 6180
6170      IF W(K1) = C0 THEN 6220
6180    NEXT K1

6190    REM LAST DATA VALUE TO BE DISPLAYED--POINT PAST IT FOR
                                                      CONSISTENCY

6200    LET K1 = A2 + 1
6210    GO TO 6290
6220    IF C0 <> 0 THEN 6290

6230    REM   ZERO CUT: IF DATA ALL <=0, ALL ZEROS GO ON "-0" STEM

6240    IF H1 <= 0 THEN 6200

6250    REM BOTH +0 AND -0 STEMS -- SHARE THE ZERO'S BETWEEN THEM
6260    REM USE COUNTS PLACED IN W(N+1) TO W(N+3)
6270    REM TO ASSIGN 'SIGNED' ZEROS PROPERLY

6280    LET K1 = K1 + W(Z0 - 1) + FNI(W(Z0) / 2)

6290    REM   COMPUTE DEPTH IN C$

6300    LET P5 = K1 - K2
6310    LET P6 = P6 + P5

6320    REM   CASE: WHERE IS THE MEDIAN?

6330    IF D0 = 0 THEN 6370

6340    REM   CASE 1: PAST THE MEDIAN

6350    LET C$ = STR$(N - (P6 - P5))
6360    GO TO 6490
6370    IF P6 <> N / 2 THEN 6410

6380    REM   CASE 2: MEDIAN BETWEEN STEMS

6390    LET D0 = 1
6400    GO TO 6470
6410    IF P6 < (N + 1) / 2 THEN 6470
```

```
6420    REM   CASE 3: MEDIAN ON THIS LINE

6430    LET C$ = STR$(P5)
6440    PRINT TAB(I8 - 6 - LEN(C$) - 1);"(";C$;")";
6450    LET D0 = 1
6460    GO TO 6500

6470    REM   CASE 4: NOT UP TO MEDIAN YET

6480    LET C$ = STR$(P6)
6490    PRINT TAB(I8 - 6 - LEN(C$));C$;

6500    REM   FIND AND PRINT LINE LABEL. L2 IS LEAF DIGIT.
6510    REM   S2 IS STEM, C$ HOLDS LABEL.

6520    LET S2 = FNI(L4 / 10)
6530    LET L2 = ABS(L4 - S2 * 10)
6540    LET C$ = STR$(S2)

6550    REM CASE: HOW MANY POSSIBLE DIGITS/LINE.
6560    REM   CONSULT THE LINE WIDTH, L9.

6570    IF L9 = 10 THEN 6890
6580    IF L9 = 5 THEN 6790

6590    REM   L9=2: 2 POSSIBLE DIGITS/LINE; 5 LINES/STEM

6600    IF S2 <> 0 THEN 6670
6610    IF L2 > 1 THEN 6670
6620    IF N7 = 0 THEN 6650
6630    PRINT TAB(I8 - 4);"-0* ";
6640    GO TO 6950
6650    PRINT TAB(I8 - 4);"+0* ";
6660    GO TO 6950

6670    REM   NOT A ZERO--PRINT LABEL

6680    ON FNI(L2 / 2) + 1 GO TO 6690,6710,6730,6750,6770
6690    PRINT TAB(I8 - LEN(C$) - 2);C$;"* ";
6700    GO TO 6950
6710    PRINT TAB(I8 - 2);"T ";
6720    GO TO 6950
6730    PRINT TAB(I8 - 2);"F ";
6740    GO TO 6950
6750    PRINT TAB(I8 - 2);"S ";
6760    GO TO 6950
6770    PRINT TAB(I8 - LEN(C$) - 2);C$;". ";
6780    GO TO 6950

6790    REM   L9=5: 5 POSSIBLE DIGITS/LINE; 2 LINES/STEM

6800    IF L2 >= 5 THEN 6870
6810    IF S2 <> 0 THEN 6850
6820    IF N7 <> 1 THEN 6850
```

```
6830    REM   "-0*" LINE -- PRINT THE "-"

6840    PRINT TAB(I8 - 3);"-";
6850    PRINT TAB(I8 - LEN(C$) - 1);C$;"* ";
6860    GO TO 6950
6870    PRINT TAB(I8 - 1);". ";
6880    GO TO 6950

6890    REM L9=10: 10 POSSIBLE DIGITS/LINE; 1 LINE/STEM

6900    IF S2 <> 0 THEN 6940
6910    IF N7 <> 1 THEN 6940
6920    PRINT TAB(I8 - 3);"-0 ";
6930    GO TO 6950
6940    PRINT TAB(I8 - LEN(C$) - 1);C$;" ";

6950    REM   FROM K2 TO K1, FIND LEAVES AND PRINT THEM. D = LEAF.

6960    IF K2 = K1 THEN 7070
6970    LET D = ABS(W(K2) - FNI(W(K2) / 10) * 10)
6980    PRINT STR$(D);
6990    LET I1 = I1 + 1
7000    IF I1 < I9 - 1 THEN 7040
7010    PRINT "*";
7020    LET K2 = K1
7030    GO TO 7050
7040    LET K2 = K2 + 1
7050    IF K2 > N THEN 7170
7060    IF K2 < K1 THEN 6970

7070    REM   END LINE

7080    PRINT
7090 NEXT J1

7100 REM   PRINT HIGH VALUES BEYOND ADJACENT VALUE

7110 IF K1 > N THEN 7170
7120 PRINT
7130 PRINT TAB(I8 - 4);"HI: ";
7140 FOR I = K1 TO N
7150    PRINT STR$(W(I));", ";
7160 NEXT I
7170 PRINT
7180 RETURN
```

```
      SUBROUTINE STMNLF(Y, N, SORTY, IW, XTREMS, ERR)
C
      INTEGER N, IW(N), ERR
      REAL Y(N), SORTY(N)
      LOGICAL XTREMS
C
C PRODUCE A STEM-AND-LEAF DISPLAY OF THE DATA IN Y()
C
C IW() IS AN INTEGER WORK ARRAY. SORTY() IS A REAL WORK ARRAY
C XTREMS IS A LOGICAL FLAG, .TRUE. IF SCALING TO EXTREMES.
C   (OTHERWISE, SCALES TO FENCES).
C
C COMMON  BLOCKS AND VARIABLES FOR OUTPUT
C
      COMMON/CHRBUF/P, PMAX, PMIN, OUTPTR, MAXPTR, OUNIT
      INTEGER P(130), PMAX, PMIN, OUTPTR, MAXPTR, OUNIT
      COMMON/NUMBRS/EPSI, MAXINT
      REAL EPSI, MAXINT
C
C FUNCTIONS
C
      INTEGER INTFN, FLOOR
C
C  CALLS SUBROUTINES DEPTHP, NPOSW, OUTLYP, PRINT, PUTCHR, PUTNUM,
C  SLTITL, STEMP, YINFO
C
C LOCAL VARIABLES
C
      REAL MED, HL, HH, ADJL, ADJH, STEP, UNIT, FRACT, NICNOS(4), NPW
      INTEGER I, SLBRK, PLTWID, RANK, IADJL, IADJH, NLINS
      INTEGER NLMAX, LINWID
      INTEGER LOW, HI, CUT, STEM, PT1, PT2, J, SPACNT, LEAF, NN, CHSTAR
      LOGICAL NEGNOW, MEDYET
C
C DATA DEFINITIONS: A USEFUL CHARACTER AND THE SCALING OPTIONS
C
      DATA CHSTAR/41/
      DATA NICNOS(1), NICNOS(2), NICNOS(3), NICNOS(4)/1.0,2.0,5.0,10.0/
      DATA NN/4/
C
C
      IF(N .GE. 2) GO TO 5
      ERR = 11
      GO TO 999
C
C SETUP -- FIND WIDTH OF PLOTTING REGION, STEM-LEAF BREAK POSITION, ETC
C
    5 SLBRK = PMIN + 11
      PLTWID = PMAX - SLBRK - 2
      IF(PLTWID .GT. 5) GO TO 10
      ERR = 13
      GO TO 999
```

27

```
C
C FIND THE BEST SCALE FOR THE PLOT
C
C SORT Y IN SORTY AND GET SUMMARY INFORMATION
C
   10 DO 20 I = 1, N
         SORTY(I) = Y(I)
   20 CONTINUE
      CALL YINFO(SORTY,N,MED,HL,HH,ADJL,ADJH,IADJL,IADJH,STEP,ERR)
      IF(ERR .NE. 0) GO TO 999
C
C FIND NICE LINE WIDTH FOR PLOT
C
C  IF ADJACENT VALUES EQUAL OR USEP DEMANDS IT, FAKE THE ADJACENT
C  VALUES TO BE THE EXTREMES
C
      IF((ADJH .GT. ADJL) .AND. .NOT. XTREMS) GO TO 25
      IADJL = 1
      IADJH = N
      ADJL = Y(IADJL)
      ADJH = Y(IADJH)
   25 NLMAX = INTFN(10.0*ALOG10(FLOAT(IADJH - IADJL + 1)), ERR)
      IF(ADJH .GT. ADJL) GO TO 27
C
C  EVEN IF ALL VALUES ARE EQUAL WE CAN PRODUCE A DISPLAY
C
      ADJH = ADJL + 1.0
      NLMAX = 1
   27 CALL NPOSW(ADJH, ADJL, NICNOS, NN, NLMAX, .TRUE., NLINS, FRACT,
     1  UNIT, NPW, ERR)
      IF(ERR .NE. 0) GO TO 999
C
C RESCALE EVERYTHING ACCORDING TO UNIT.  HEREAFTER EVERYTHING IS
C INTEGER, AND DATA ARE OF THE FORM     SS...SL(.)
C NOTE THAT INTFN PERFORMS EPSILON ADJUSTMENTS FOR CORRECT ROUNDING,
C AND CHECKS THAT THE REAL NUMBER IS NOT TOO LARGE FOR AN INTEGER
C VARIABLE.
C
      DO 30 I = 1, N
         IW(I) = INTFN(SORTY(I)/UNIT, ERR)
   30 CONTINUE
      IF(ERR .NE. 0) GO TO 999
C
      IF (FRACT .EQ. 10.0) GO TO 40
C
C IF ALL LEAVES ARE ZERO, WE SHOULD BE IN ONE-LINE-PER-STEM FORMAT
C
      DO 35 I = IADJL, IADJH
         IF (MOD(IW(I), 10) .NE. 0) GO TO 40
   35 CONTINUE
```

```
      FRACT = 10.0
      NPW = FRACT * UNIT
      NLINS = INTFN(ADJH/NPW, ERR) - INTFN(ADJL/NPW, ERR) + 1
      IF(ADJH * ADJL .LT. 0.0  .OR.  ADJH .EQ. 0.0) NLINS = NLINS+1
   40 LOW = IW(IADJL)
      HI = IW(IADJH)
C
C LINEWIDTH NOW IS NICEWIDTH/UNIT = FRACT
C
      LINWID = INTFN(FRACT, ERR)
C
      CALL SLTITL(UNIT, ERR)
      IF(ERR .NE. 0) GO TO 999
C
C PRINT VALUES BELOW LOW ADJACENT VALUE ON "LO" STEM
C
      RANK = IADJL - 1
      IF(IADJL .EQ. 1) GO TO 50
      CALL OUTLYP(IW, N, 1, RANK, .FALSE., SLBRK, ERR)
      IF(ERR .NE. 0) GO TO 999
C
C INITIALIZE FOR MAIN PART OF DISPLAY.
C INITIAL SETTINGS ARE TO LINE BEFORE FIRST ONE PRINTED
C
   50 CUT = FLOOR((1.0 + EPSI)*FLOAT(LOW)/FLOAT(LINWID)) * LINWID
      NEGNOW = .TRUE.
      STEM = CUT
      IF(LOW .LT. 0) GO TO 60
C
C FIRST STEM POSITIVE
C
      NEGNOW = .FALSE.
      STEM = CUT - LINWID
   60 MEDYET = .FALSE.
C
C TWO POINTERS ARE USED.  PT1 COUNTS FIRST FOR DEPTHS, PT2 FOLLOWS
C FOR LEAF PRINTING.  BOTH ARE INITIALIZED ONE POINT EARLY.
C
      PT1 = IADJL
      PT2 = PT1
C
C MAIN LOOP.  FOR EACH LINE
C
      DO 120 J = 1, NLINS
C VARIABLE USES:
C  CUT =      FIRST NUMBER ON NEXT LINE OF POSITIVE STEMS, BUT
C      =      LAST NUMBER ON CURRENT LINE OF NEGATIVE STEMS
C  STEM =     INNER (NEAR ZERO) EDGE OF CURRENT LINE
C  SPACNT     COUNTS SPACES USED ON THIS LINE
```

```
C
C STEP TO NEXT LINE
      CUT = CUT + LINWID
C
C IF(STEM = 0 AND NEGNOW) NEGNOW  =  .F. ELSE STEM = STEM + LINWID
C
      IF(STEM .NE. 0 .OR. .NOT. NEGNOW) GO TO 70
      NEGNOW = .FALSE.
      GO TO 80
   70 STEM = STEM + LINWID
C
C NEWLINE -- INITIALIZE COUNT OF SPACES USED
C
   80 SPACNT = 0
C
C FIND AND PRINT DEPTH
C
      CALL DEPTHP(SORTY, IW, N, PT1, PT2, CUT, IADJH, HI, RANK,
     1    MEDYET, SLBRK, ERR)
      IF(ERR .NE. 0) GO TO 999
C
C PRINT STEM LABEL
C
      CALL STEMP(STEM, LINWID, NEGNOW, SLBRK, ERR)
      IF(ERR .NE. 0) GO TO 999
C
C FIND AND PRINT LEAVES
C
      IF (PT1 .EQ. PT2) GO TO 110
   90 LEAF = IABS(IW(PT2) - (STEM/10)*10)
      CALL PUTNUM(0, LEAF, 1, ERR)
      SPACNT = SPACNT + 1
      IF(SPACNT .LT. PLTWID) GO TO 100
C
C LINE OVERFLOWS PAST RIGHT EDGE.   MARK WITH *
C
      CALL PUTCHR(0, CHSTAR, ERR)
      IF(ERR .NE. 0) GO TO 999
      PT2 = PT1
      GO TO 110
  100 PT2 = PT2 + 1
      IF(PT2 .LT. PT1) GO TO 90
C
C END LINE
C
  110 CALL PRINT
C
C CONTINUE LOOP UNTIL WE RUN OUT OF  NUMBERS TO PLOT
C
  120 CONTINUE
```

```
C
C   PRINT VALUES ABOVE HI ADJACENT VALUE ON "HI" STEM
C
        IF(PT1 .GT. N) GO TO 990
        CALL OUTLYP(IW, N, PT1, N, .TRUE., SLBRK, ERR)
   990 WRITE(OUNIT, 5990)
  5990 FORMAT(1X)
   999 RETURN
        END


        SUBROUTINE OUTLYP(IW, N, FROM, TO, HIEND, SLBRK, ERR)
C
        LOGICAL HIEND
        INTEGER N, IW(N), FROM, TO, SLBRK, ERR
C
C   PRINT THE LO OR HI STEM FOR A STEM-AND-LEAF DISPLAY.
C   THE LOGICAL VARIABLE  HIEND  IS  .TRUE.  IF WE ARE TO PRINT
C   THE  HI  STEM, .FALSE.  IF THE  LO  STEM IS TO BE PRINTED.
C   IW()  CONTAINS  N  SORTED AND SCALED DATA VALUES.  EACH HAS THE
C     FORM SS...SL, WHERE THE ONE'S DIGIT IS THE LEAF.
C   FROM, TO  ARE POINTER INTO  IW() DELIMITING THE VALUES TO BE
C     PLACED ON THE  HI  OR  LO  STEM.
C   SLBRK  IS THE CHARACTER POSITION ON THE PAGE OF THE BLANK COLUMN
C     BETWEEN STEMS AND LEAVES.
C
C
C
C COMMON FOR OUTPUT
C
        COMMON /CHRBUF/P, PMAX, PMIN, OUTPTR, MAXPTR, OUNIT
        INTEGER P(130), PMAX, PMIN, OUTPTR, MAXPTR, OUNIT
C
C FUNCTIONS
C
        INTEGER WDTHOF
C
C LOCAL VARIABLES
C
        INTEGER CHL, CHO, CHH, CHI, CHCOMA, CHBL, OPOS, NWID, LHMAX, I
C
C NEEDED CHARACTERS
C
        DATA CHH, CHI, CHL, CHO, CHCOMA, CHBL/8, 9, 12, 15, 45, 37/
C
        OPOS = SLBRK - 3
        IF(HIEND) GO TO 10
        CALL PUTCHR(OPOS, CHL, ERR)
        CALL PUTCHR(0, CHO, ERR)
        GO TO 20
    10 CALL PRINT
        CALL PUTCHR(OPOS, CHH, ERR)
        CALL PUTCHR(0, CHI, ERR)
    20 CALL PUTCHR(SLBRK, CHBL, ERR)
        IF(ERR .NE. 0) GO TO 999
```

```
      NWID = MAXO( WDTHOF(IW(FROM)), WDTHOF(IW(TO)) )
      LHMAX = PMAX - NWID - 2
      DO 40 I = FROM, TO
        CALL PUTNUM(O, IW(I), NWID, ERR)
        CALL PUTCHR(O, CHCOMA, ERR)
        CALL PUTCHR(O, CHBL, ERR)
        IF(OUTPTR .LT. LHMAX) GO TO 30
        CALL PRINT
        CALL PUTCHR(SLBRK, CHBL, ERR)
   30   IF(ERR .NE. 0) GO TO 999
   40 CONTINUE
C
C BUT DONT PRINT THE FINAL COMMA
C
      OPOS = MAXPTR - 1
      CALL PUTCHR(OPOS, CHBL, ERR)
      CALL PRINT
      IF(.NOT. HIEND) CALL PRINT
  999 RETURN
      END


      SUBROUTINE DEPTHP(W, IW, N, PT1, PT2, CUT, IADJH, HI, RANK,
     1 MEDYET, SLBRK, ERR)
C
C COMPUTE AND PRINT THE DEPTH FOR THE CURRENT LINE
C
      LOGICAL MEDYET
      INTEGER N, PT1, PT2, CUT, IADJH, HI, RANK, SLBRK, ERR
      INTEGER IW(N)
      REAL W(N)
C
C  W()   HOLDS THE  N  SORTED DATA VALTUES
C  IW()   HOLDS THE SCALED VERSION OF W()
C  PT1, PT2  ARE POINTERS INTO IW() AND W().  ON ENTRY,
C         PT1 = PT2 POINT TO THE FIRST DATA VALUE NOT YET PRINTED.
C        ON EXIT, PT1 POINTS TO THE FIRST DATA VALUE ON THE NEXT LINE,
C         PT2 IS UNCHANGED.
C  CUT    THE LARGEST VALUE ON THE CURRENT (POSITIVE) LINE, OR THE
C               SMALLEST VALUE ABOVE THE CURRENT (NEGATIVE) LINE.
C  IADJH  POINTS TO THE HIGH ADJACENT VALUE IN W() AND IW()
C  HI   IS THE GREATEST VALUE BEING DISPLAYED
C  RANK   A RUNNING TOTAL OF THE RANK FROM THE LOW END.  ON EXIT,
C        RANK  IS UPDATED TO INCLUDE THE COUNT FOR THE CURRENT LINE.
C  MEDYET  IS A LOGICAL FLAG, SET  .TRUE.  WHEN THE MEDIAN VALUE HAS
C               BEEN PROCESSED.
C  SLBRK  IS THE CHARACTER POSITION ON THE PAGE OF THE BLANK COLUMN
C               BETWEEN THE STEMS AND LEAVES.
C
```

```
C
C FUNCTIONS
C
      INTEGER INTFN, WDTHOF
C
C LOCAL VARIABLES
C
      INTEGER CHLPAR, CHRPAR, LEFCNT, PTZ, DEPTH, NWID, OPOS, PTX
C
C OUTPUT CONTROL
C
      COMMON/CHRBUF/P, PMAX, PMIN, OUTPTR, MAXPTR, OUNIT
      INTEGER P(130), PMAX, PMIN, OUTPTR, MAXPTR, OUNIT
      DATA CHLPAR, CHRPAR/43, 44/
C
      PTX = PT1
      DO 90 PT1 = PTX, IADJH
        IF(IW(PT1) .GT. CUT) GO TO 110
        IF((CUT .GE. 0) .AND. (IW(PT1) .EQ. CUT)) GO TO 110
   90 CONTINUE
C
C  LAST DATA VALUE IF WE FALL THRU HERE--POINT PAST IT FOR CONSISTENCY.
C
  100 PT1 = IADJH+1
      GO TO 140
  110 IF(CUT .NE. 0) GO TO 140
C
C  ZERO CUT:  IF DATA ALL .LE. 0, ALL ZEROES GO ON "-0" STEM
C
      IF(HI .LE. 0) GO TO 100


C
C  BOTH +0 AND -0 STEMS -- SHARE THE ZEROES BETWEEN THEM
C
C FIRST CHECK FOR NUMBERS ROUNDED TO ZERO--TRUE -OS
      DO 115 PTZ = PT1, N
        IF(W(PTZ).GE. 0.0) GO TO 117
  115 CONTINUE
  117 PT1 = PTZ
      DO 120 PTZ = PT1, N
        IF(W(PTZ) .GT. 0.0) GO TO 130
  120 CONTINUE
  130 PT1 = PT1 + INTFN(FLOAT(PTZ - PT1)/2.0, ERR)
C
C  COMPUTE AND PRINT DEPTH
C
  140 LEFCNT = PT1 - PT2
      RANK = RANK + LEFCNT
C
```

```
C
C   CASE: WHERE IS THE MEDIAN?
C
C
        IF(.NOT. MEDYET) GO TO 150
C
C    CASE 1: PAST THE MEDIAN
C
        DEPTH = N - (RANK - LEFCNT)
        GO TO 180
   150 IF(FLOAT(RANK) .NE. FLOAT(N)/2.0) GO TO 160
C
C    CASE 2: MEDIAN FALLS BETWEEN STEMS AT THIS POINT
C
        MEDYET = .TRUE.
        GO TO 170
   160 IF( FLOAT(RANK) .LT. FLOAT(N+1)/2.0) GO TO 170
C
C    CASE 3: MEDIAN IS ON THE CURRENT LINE
C
        NWID = WDTHOF(LEFCNT)
        OPOS = SLBRK - 7 - NWID
        CALL PUTCHR(OPOS, CHLPAR, ERR)
        CALL PUTNUM(0, LEFCNT, NWID, ERR)
        CALL PUTCHR(0, CHRPAR, ERR)
        MEDYET = .TRUE.
        GO TO 999
C
C    CASE 4: NOT UP TO MEDIAN YET
C
   170 DEPTH = RANK
C
C   PRINT THE DEPTH, IF IT HASN'T BEEN DONE YET
C
   180 NWID = WDTHOF(DEPTH)
        OPOS = SLBRK - 6 - NWID
        CALL PUTNUM(OPOS, DEPTH, NWID, ERR)
   999 RETURN
        END
```

```
          SUBROUTINE STEMP(STEM, LINWID, NEGNOW, SLBRK, ERR)
C
C   COMPUTE AND "PRINT" THE STEM
C
          LOGICAL NEGNOW
          INTEGER STEM, LINWID, SLBRK, ERR
C
C   ON ENTRY:
C    STEM  IS THE INNER (NEAR ZERO) EDGE OF THE CURRENT LINE
C    LINWID  IS THE NUMBER OF POSSIBLE DIFFERENT LEAF DIGITS
C    NEGNOW  IS .TRUE. IF THE CURRENT LINE IS NEGATIVE
C    SLBRK  IS THE CHARACTER POSITION ON THE PAGE OF THE BLANK
C         COLUMN BETWEEN STEMS AND LEAVES
C
C
C   COMMONS FOR OUTPUT
C
          COMMON /CHRBUF/P, PMAX, PMIN, OUTPTR, MAXPTR, OUNIT
          INTEGER P(130), PMAX, PMIN, OUTPTR, MAXPTR, OUNIT
C
C   FUNCTION
C
          INTEGER WDTHOF
C
C   LOCAL VARIABLES
C
          INTEGER CH0, CHBL, CHPLUS, CHMIN, CHSTAR, CHPT
          INTEGER NSTEM, LEFDIG, NWID, OPOS, OCHR, I, CH5STM(5)
          DATA CH0/27/
          DATA CHBL, CHPLUS, CHMIN, CHSTAR, CHPT/37, 39, 40, 41, 46/
          DATA CH5STM(1),CH5STM(2),CH5STM(3),CH5STM(4)/41,20,6,19/
          DATA CH5STM(5)/46/
C
          NSTEM = STEM/10
          LEFDIG = IABS(STEM - NSTEM * 10)
          NWID = WDTHOF(NSTEM)
C
C
C   CASE: HOW MANY POSSIBLE DIGITS/LINE ( = LINWID)
C
C
          IF(LINWID .NE. 2) GO TO 260
```

```
C
C    CASE 1: 2 POSSIBLE DIGITS/LINE; 5 LINES/STEM
C
      IF(NSTEM .NE. 0) GO TO 200
C    PLUS OR MINUS ZERO
      OPOS = SLBRK - 4
      IF(NEGNOW) CALL PUTCHR(OPOS, CHMIN, ERR)
      IF(.NOT. NEGNOW) CALL PUTCHR(OPOS, CHPLUS, ERR)
      OPOS = OPOS + 1
      GO TO 210
  200 OPOS = SLBRK - NWID - 2
  210 CALL PUTNUM(OPOS, NSTEM, NWID, ERR)
      I = LEFDIG/2 + 1
      OCHR = CH5STM(I)
      CALL PUTCHR(0, OCHR, ERR)
      GO TO 990
  260 IF(LINWID .NE. 5) GO TO 290
C
C    CASE 2: 5 POSSIBLE DIGITS/LINE; 2 LINES/STEM
C
      OPOS = SLBRK - NWID - 1
      IF(NSTEM .NE. 0) GO TO 270
C
C    -0*    PRINT THE SIGN (IT APPEARS AUTOMATICALLY OTHERWISE)
C
      OPOS = SLBRK - 3
      IF(NEGNOW) CALL PUTCHR(OPOS, CHMIN, ERR)
      IF(.NOT. NEGNOW) CALL PUTCHR(OPOS, CHPLUS, ERR)
  270 OPOS = SLBRK - NWID - 1
      CALL PUTNUM(OPOS,NSTEM,NWID,ERR)
      IF(LEFDIG .LT. 5) CALL PUTCHR(0,CHSTAR,ERR)
      IF(LEFDIG .GE. 5) CALL PUTCHR(0,CHPT,ERR)
      GO TO 990
C
C    CASE 3: 10 POSSIBLE DIGITS/LEAF; 1 LINE/STEM
C
  290 IF(LINWID .EQ. 10) GO TO 300
C
C    ILLEGAL VALUE -- NICE NUMBERS BAD?
C
      ERR = 12
      GO TO 999
  300 IF((NSTEM .NE. 0) .OR. .NOT. NEGNOW) GO TO 310
      OPOS = SLBRK - 3
      CALL PUTCHR(OPOS,CHMIN,ERR)
      CALL PUTCHR(0,CHO,ERR)
      GO TO 990
  310 OPOS = SLBRK - NWID -1
      CALL PUTNUM(OPOS,NSTEM,NWID,ERR)
  990 CALL PUTCHR(SLBRK,CHBL,ERR)
  999 RETURN
      END
```

```
      SUBROUTINE SLTITL (UNIT, ERR)
C
C     PRINT THE TITLE FOR A STEM-AND-LEAF DISPLAY
C
      INTEGER ERR
      REAL UNIT
C
C     ON ENTRY:
C      UNIT  IS THE LEAF DIGIT UNIT
C
C     NOTE THAT THIS ROUTINE CAN BE MODIFIED TO PRINT THE NAME OF
C     THE BATCH BEING DISPLAYED IF SUCH A NAME IS KNOWN.
C
C
C
C     COMMON BLOCKS
C
      COMMON /CHARIO/ CHARS, CMAX,
     1 CHA, CHB, CHC, CHD, CHE, CHF, CHG, CHH, CHI, CHJ, CHK,
     2 CHL, CHM, CHN, CHO, CHP, CHQ, CHR, CHS, CHT, CHU, CHV,
     3 CHW, CHX, CHY, CHZ, CH0, CH1, CH2, CH3, CH4, CH5, CH6,
     4 CH7, CH8, CH9, CHBL, CHEQ, CHPLUS, CHMIN, CHSTAR, CHSLSH,
     5 CHLPAR, CHRPAR, CHCOMA, CHPT
      COMMON/CHRBUF/P, PMAX, PMIN, OUTPTR, MAXPTR, OUNIT
      INTEGER P(130), PMAX, PMIN, OUTPTR, MAXPTR, OUNIT
      INTEGER CHARS(46), CMAX
      INTEGER CHA, CHB, CHC, CHD, CHE, CHF, CHG, CHH, CHI, CHJ, CHK
      INTEGER CHL, CHM, CHN, CHO, CHP, CHQ, CHR, CHS, CHT, CHU, CHV
      INTEGER CHW, CHX, CHY, CHZ, CH0, CH1, CH2, CH3, CH4, CH5, CH6
      INTEGER CH7, CH8, CH9, CHBL, CHEQ, CHPLUS, CHMIN, CHSTAR, CHSLSH
      INTEGER CHLPAR, CHRPAR, CHCOMA, CHPT
C
C     FUNCTIONS
C
      INTEGER INTFN, WDTHOF
C
C     LOCAL VARIABLES
C
      INTEGER IEXPT, OWID, NUM, I
C
      WRITE(OUNIT, 5000) UNIT
 5000 FORMAT(24H    STEM-AND-LEAF DISPLAY/20H    LEAF DIGIT UNIT =, F9.4)
```

```
C
C   PRINT "  1  2   REPRESENTS "
C
      CALL PUTCHR(0, CHBL, ERR)
      CALL PUTCHR(0, CHBL, ERR)
      CALL PUTCHR(0, CH1, ERR)
      CALL PUTCHR(0, CHBL, ERR)
      CALL PUTCHR(0, CHBL, ERR)
      CALL PUTCHR(0, CH2, ERR)
      CALL PUTCHR(0, CHBL, ERR)
      CALL PUTCHR(0, CHBL, ERR)
      CALL PUTCHR(0, CHR, ERR)
      CALL PUTCHR(0, CHE, ERR)
      CALL PUTCHR(0, CHP, ERR)
      CALL PUTCHR(0, CHR, ERR)
      CALL PUTCHR(0, CHE, ERR)
      CALL PUTCHR(0, CHS, ERR)
      CALL PUTCHR(0, CHE, ERR)
      CALL PUTCHR(0, CHN, ERR)
      CALL PUTCHR(0, CHT, ERR)
      CALL PUTCHR(0, CHS, ERR)
      CALL PUTCHR(0, CHBL, ERR)
C
C   AND FINISH IT OFF
C
      IEXPT = INTFN(ALOG10(UNIT),ERR)
      IF(IEXPT .GE. 0) GO TO 200
      IF(IEXPT .EQ. (-1)) GO TO 100
C
C UNIT .LE. 0.01
C
      IEXPT = IABS(IEXPT) - 2
      CALL PUTCHR(0, CH0, ERR)
      CALL PUTCHR(0, CHPT, ERR)
      IF(IEXPT .EQ. 0) GO TO 30
      DO 20 I = 1, IEXPT
        CALL PUTCHR(0, CH0, ERR)
   20 CONTINUE
   30 CALL PUTCHR(0, CH1, ERR)
      CALL PUTCHR(0, CH2, ERR)
      GO TO 900
C
C   PRINT 1.2
C
  100 CALL PUTCHR(0, CH1, ERR)
      CALL PUTCHR(0, CHPT, ERR)
      CALL PUTCHR(0, CH2, ERR)
      GO TO 900
```

```
C
C   UNIT .GE. 1.0
C
  200 NUM = 12 * INTFN(UNIT,ERR)
      OWID = WDTHOF(NUM)
      CALL PUTNUM(O, NUM, OWID, ERR)
      CALL PUTCHR(O, CHPT, ERR)
C
C   WRAP UP
C
  900 IF (ERR .NE. 0) GO TO 999
      CALL PRINT
      WRITE(OUNIT, 5010)
 5010 FORMAT(/)
  999 RETURN
      END
```

Chapter 2 _____

Letter-Value Displays

It is often convenient to summarize a data batch after we have taken an initial look at it and have seen each individual data item. For example, we can use a central value to summarize the size or general level of the numbers in the batch. We also want to describe how spread out or variable the numbers in the batch are, and we might look for ways to describe more precisely the shapes and patterns we can see in the outline of a stem-and-leaf display. As always, when we explore data, we must be alert for extraordinary values that might require special attention. Letter values provide information for several of these summaries, and the letter-value display presents the letter values in a convenient form.

2.1   Median, Hinges, and Other Summary Values

Before we determine the letter values, we must first order the data batch from lowest value to highest. When we analyze data by hand, a stem-and-leaf display provides a quick, crude ordering of the batch. Computers can order the data with special sorting programs (see Section 2.9). When a data batch is ordered, a set of suitably selected data values and simple averages of these values can convey many important features of the batch concisely. The letter values are just such a set of values.

depth

One of the most important characteristics of a data value in an ordered batch is how far it is from the low or high end of the batch. We therefore define the *depth* of each data value. This is just the value's position in an enumeration of values that starts at the nearer end of the batch. (Recall from Chapter 1 that depths appear in a column at the left of a finished stem-and-leaf display.) Each extreme value is the first value in the enumeration and therefore has a depth of 1; the second largest and second smallest values each have a depth of 2; and so on. In general, in a batch of size n, two data values have depth i: the ith and the $(n + 1 - i)$th. Conversely, the depth of the ith data value in an ordered batch is the smaller of i and $n + 1 - i$ because depth is measured from the nearer end. We find letter values at certain selected depths.

median

If n is odd, there is a "deepest" data value—one as far from either end of the ordered batch as possible, and thus not part of a pair of equal-depth numbers. This data value is the *median,* and it marks the middle of the batch—in the sense that exactly half the remaining $n - 1$ numbers in the batch are less than or equal to it, and exactly half are greater than or equal to it.

It is easy to calculate the depth of the median. It is simply $(n + 1)/2$. Because the depth of the ith data value in an ordered batch of n values is the smaller of i and $n + 1 - i$, the maximum depth occurs where $i = n + 1 - i$, or, equivalently, $2i = n + 1$. Thus

$$\text{depth of median} = (n + 1)/2,$$

which we abbreviate $d(M) = (n + 1)/2$. For example, if we have 3 data values in order, the median is the second value because one value is less than the median and one is greater. If we have a batch of 5 values, the median has 2 values below it and 2 values above it, so that it is the third largest value or the third smallest value, depending on whether we count from the top or from the bottom.

But, what if a batch has an even count? We then have two "middle" values. If these two values are different—as they usually are—no one data value divides the batch in half. Then, $d(M) = (n + 1)/2$ will have a fractional part equal to $\frac{1}{2}$, and this depth points between the two middle data values. Because half the data values lie below the median and half lie above it, we adopt the usual convention of averaging the middle two data values, each of which has depth $(n + 1)/2 - \frac{1}{2}$. We label the median with the letter M.

The median splits an ordered batch in half. We might naturally ask
hinges next about the middle of each of these halves. The *hinges* are the summary
values in the middle of each half of the data. They are denoted by the letter H
and are about a quarter of the way in from each end of the ordered batch. We
find hinges in much the same way as we found the median. We begin with
$d(M)$, the depth of the median, drop off the fraction of ½ if there is one, add 1,
and find

$$d(H) = ([d(M)] + 1)/2,$$

where the [ ] symbols are read "integer part of" and indicate the operation of
omitting the fraction. Each hinge is at depth $d(H)$, and again a fraction of ½
tells us to average the two data values surrounding that depth.

quartiles The hinges are similar to the *quartiles,* which are defined so that one
quarter of the data lies below the lower quartile and one quarter of the data
lies above the upper quartile.* The main difference between hinges and
quartiles is that the depth of the hinges is calculated from the depth of the
median, with the result that the hinges often lie slightly closer to the median
than do the quartiles. This difference is quite small, and the arithmetic
required to calculate the depth of the hinges is simpler.

The next step is almost automatic. We find middle values for the outer
quarters of the data. These values are about an eighth of the way in from each
eighths end of the ordered batch. They are called *eighths* and are denoted by the letter
E. Their depth is

$$d(E) = ([d(H)] + 1)/2,$$

where, again, the [ ] symbols tell us to drop any fraction in $d(H)$, and a new
fraction of ½ tells us to average adjacent data values.

Example: New Jersey Counties

Exhibit 2–1 lists the area in square miles of the 21 counties of New Jersey.
Sorted into increasing order, the areas are 47, 103, 130, 192, 221, 228, 234,
267, 307, 312, 329, 362, 365, 423, 468, 476, 500, 527, 569, 642, 819. Here
$n = 21$, and

$$d(M) = ([21] + 1)/2 = 11.$$

---

*Hinges are sometimes called *quarters* or *fourths*. The latter term may well replace *hinges* in time, but this
book uses the term *hinges* for compatibility with *Exploratory Data Analysis*.

Exhibit 2–1   Area of New Jersey Counties (in square miles)

Atlantic	569	Middlesex	312
Bergen	234	Monmouth	476
Burlington	819	Morris	468
Camden	221	Ocean	642
Cape May	267	Passaic	192
Cumberland	500	Salem	365
Essex	130	Somerset	307
Gloucester	329	Sussex	527
Hudson	47	Union	103
Hunterdon	423	Warren	362
Mercer	228		

Source: U.S. Bureau of the Census, *County and City Data Book, 1977* (Washington, D.C.: Government Printing Office, 1978).

The eleventh value, if we count from either end, is 329; this value is the median.

Since $d(M) = 11$, the depth of the hinge is

$$d(H) = ([d(M)] + 1)/2 = 12/2 = 6.$$

Thus, the two hinges are 228, the sixth value from the bottom, and 476, the sixth value from the top. Then, the depth of the eighths is

$$d(E) = ([d(H)] + 1)/2 = (6 + 1)/2 = 3\tfrac{1}{2}.$$

Thus the two eighths are found by averaging the third and fourth values from each end: $(130 + 192)/2 = 161$ and $(569 + 527)/2 = 548$.

2.2   Letter Values

letter values

The summary values we have been examining—the median, the hinges, and the eighths—are the start of the sequence of *letter values,* so called because we often label them with single letters—M, H, and E. The letter values beyond the eighths are used less frequently. Generally, these values are not named and are referred to by their labels—D, C, B, A, Z, Y, X, W, and so on. The depths corresponding to these labels are defined in just the same way that has taken us from median to hinge to eighth. Each subsequent depth lies halfway between the previous depth and 1, the depth of the extreme; thus, the next letter values after the eighths are labeled D and are found at depth

$$d(D) = ([d(E)] + 1)/2.$$

We continue the process of identifying letter values until we obtain a depth equal to 1. The extreme values of the batch have no letter label; they are labeled with only their depth, 1.

As we approach the extremes, we may find letter values at depth 2. When this happens, we omit the letter values at depth 1.5 ((2 + 1)/2 = 1.5)

Exhibit 2–2   Locating and Calculating the Letter Values for the New Jersey County Areas

Depths of Letter Values	Depth	Data Value	Letter Values
	1	47	extreme = 47
$d(D) = (3 + 1)/2 = 2$	2	103	D = 103
$d(E) = (6 + 1)/2 = 3.5$	3	130	E = 161
	4	192	
	5	221	
$d(H) = (11 + 1)/2 = 6$	6	228	H = 228
	7	234	
	8	267	
	9	307	
	10	312	
$d(M) = (21 + 1)/2 = 11$	11	329	M = 329
	10	362	
	9	365	
	8	423	
	7	468	
$d(H)$	6	476	H = 476
	5	500	
$d(E)$	4	527	E = 548
	3	569	
$d(D)$	2	642	D = 642
	1	819	extreme = 819

and report the extremes next. This is reasonable because the unreported letter values would just be the averages of the letter values at depths 1 and 2, which we are already reporting.

Exhibit 2–2 illustrates the connections among the data values, the depths, and the letter values.

2.3   Displaying the Letter Values

After we have determined the letter values for a batch, we need to present them in a format that helps us to see what is happening in the data. At each depth (except at the median) we have found two letter values, one by counting up toward the middle from the low end and one by counting down toward the middle from the high end. A letter-value display takes advantage of this pairing, as shown in Exhibit 2–3. In addition to the letter values and their depths, the letter-value display includes two columns of descriptive numbers, labeled "mid" and "spread." These columns provide information about the shape of the batch, as we shall soon discover.

The first two columns of Exhibit 2–3 contain the labels—M for median, H for hinge, and so on—and the depths. The columns labeled "lower" and "upper" give the lower letter values and the upper letter values respectively, with the two letter values of a pair on each line. Because the median lies at the middle of the batch and is unpaired, it straddles these two columns. The columns labeled "mid" and "spread" contain the midsummaries and the

Exhibit 2–3   Letter-Value Display for the Area of New Jersey Counties (in square miles) Shown in Exhibit 2–1

		Lower		Upper	Mid	Spread
n = 21						
M	11		329		329	
H	6	228		476	352	248
E	3.5	161		548	354.5	387
D	2	103		642	372.5	539
	1	47		819	433	772

spreads, each of which is calculated from the corresponding letter values as described in the following discussion.

In Exhibit 2–3, we readily see that the median county size is 329 square miles, that the counties range from 47 square miles to 819 square miles, and that the middle half of the 21 counties runs from 228 to 476 square miles.

midsummary
midhinge
mideighth
midextreme
midrange

Since letter values come in pairs symmetrically placed at the same depth, we might ask whether their values are also symmetric. We can find out by calculating the average value for each pair of letter values. This value midway between the two letter values is a *midsummary*. Specifically, the average of the two hinges is called the *midhinge* (midH). We can also find the *mideighth* (midE), the midD, and other midsummaries, including the *midextreme,* also called *midrange.* The median is, by being in the middle of the batch, already a midsummary. Note that, in finding midsummaries, we do not average depths, but rather we average the two letter values found at a particular depth.

We can learn a lot about how nearly symmetric a batch of values is by comparing the other midsummaries to the median or by looking for a trend in the midsummaries. If all the midsummaries are approximately equal, then the values of the hinges, eighths, and so on are nearly symmetric about the median. If the midsummaries become progressively larger, the batch is skewed toward the high side. If they decrease steadily, the batch is skewed toward the low side.

Returning to the example of the county areas, we see in Exhibit 2–3 that the midsummaries increase gradually, indicating a slight skewness toward the high side; the midextreme, 433, stands out because of the size of Burlington County.

As we noted in Chapter 1, symmetric batches of data values are often easier to summarize and analyze than batches that are asymmetric. When a batch of values is not symmetric but has a main hump and a generally smooth stem-and-leaf display, symmetry can often be attained by re-expressing the numbers. Re-expression is discussed in Section 2.4, and its use to promote symmetry is illustrated in Section 2.5.

spread

We can learn in detail how variable the data are by examining the column of spreads in a letter-value display. Each *spread* is the difference between the two letter values in a pair, calculated by subtracting the lower letter value from the upper letter value. It is named after the letter-value pair.

H-spread

For example, the *H-spread* (H-spr for short) is the difference between the hinges and thus tells the range covered by the middle half of the data. Other spreads have similar interpretations; for example, the E-spread gives the range of the middle three-quarters of the data. The difference between the extremes

range

is simply called the *range.* All these spreads respond to variability in data. The more variable the data, the larger the spreads will be. Taken together, the spreads in a letter-value display provide information about how the tails of the data behave. Section 2.6 discusses this further.

2.4   Re-expression and the Ladder of Powers

data re-expression
One way to change the shape of a batch is to *re-express* each data value in the batch. For example, we might raise each value to some power, *p*. When we work by hand, we can use a calculator or a book of tables to re-express values, but for a large batch even using a calculator can be tedious. Re-expressions are more practical when we work on a computer because the machine can do all the work quickly. When we use powers, each value of *p* will have a slightly different effect on the batch, but if we place these powers in order, their effects on the batch will also be ordered. This order leads to the ladder of powers listed in Exhibit 2–4.

The arrow in Exhibit 2–4 marks the power $p = 1$. This is "home base" because the original data values can be thought of as being re-expressed to the power 1. Raising each value in the batch to a power less than 1 will pull in a stretched-out upper tail while stretching out a bunched-in lower tail. Raising each data value to a power higher than 1 will have the reverse effect: Asymmetry to the low side will be alleviated. Thus a trend in the midsummaries indicates the direction we should move on the ladder of powers. The ladder is useful because the further we move from $p = 1$ in either direction, the greater the effect on the shape of the batch. We can thus hunt for an optimal re-expression by trying a power and examining the midsummaries in the letter-value display of the re-expressed batch. A trend in the new midsummaries will point the direction in which we should now move from where we are on the ladder for a better result. See Section 2.5 for an example.

Usually y^0 is defined to be 1. However, it would be useless to re-express all the values in a batch to 1. It turns out that, when we order the powers according to the strength of their effect on the data, the logarithm, or log, falls naturally at the zero power. The mathematical reasons for this are beyond the scope of this book, but the truth of the statement will become evident as we use the ladder of powers to find re-expressions for data.

We can save much time when working by hand by noting that we need not re-express the entire batch to construct a new letter-value display. Instead

Exhibit 2–4   Re-expressions in the Ladder of Powers $(y \rightarrow y^p)$

p	Re-expression	Name	Notes
	·		Higher powers can be used.
	·		
	·		
3	y^3	Cube	
2	y^2	Square	The highest commonly used power.
1	y^1	"Raw"	No re-expression at all.
½	$\sqrt{y}$	Square root	A commonly used power, especially for counts.
(0)	$\log(y)$	Logarithm*	$\log(y)$ holds the place of the zero power in the ladder of powers. A very common re-expression.
$-\frac{1}{2}$	$-1/\sqrt{y}$	Reciprocal root	The minus sign preserves order.
-1	$-1/y$	Reciprocal	
-2	$-1/y^2$	Reciprocal square	
	·		
	·		
	·		Lower powers can be used.

*We ordinarily use logarithms to the base 10.

we can take a shortcut and just re-express the letter values themselves or, when a depth involves ½, the two data values on which the letter value is based. Then we can compute new mids and spreads.

This shortcut is possible because every power in the ladder of powers preserves order—that is, if a is greater than b (written $a > b$) and both are positive, then $a^p > b^p$ for any non-negative power p, and $-a^p > -b^p$ for any negative p. (This is the reason for the minus signs associated with negative powers in Exhibit 2–4.) If a or b is negative, powers will not preserve order because even powers will make a^p positive, and fractional powers and the log may not even be possible. For example, $\sqrt{-2}$ and $\log(-3)$ cannot be found. Letter values are determined entirely by their depth in the ordered batch. Since the ordering of these values is not disturbed by re-expressions in the ladder of powers, the depth of every data value and the identities of the points selected as letter values remain the same. Thus we need only re-express the data values that are involved in letter values.

To streamline the process further, we could simply re-express the letter values and thus save a little effort on letter values that are the average of two data values. In general, the re-expression of an average of two data values is not identical to the average of the re-expressed data values. The difference is often slight, but not guaranteed so, especially for the more extreme letter values. The examples in this chapter do not use this shortcut.

When the numbers in a data batch are not all positive, some of the re-expressions in the ladder of powers may be impossible. For example, we cannot re-express zero by logarithms or any negative power. One way to deal *start* with this particular problem is to add a small number, or *start,* to each value in the batch before re-expressing. Thus, we might find $\log(y + \frac{1}{6})$. The value of the start usually matters little, provided it is small compared to the typical size of the data values. Starts of $\frac{1}{6}$, $\frac{1}{2}$, and 1 are commonly used.

However, we should not generally re-express negative numbers by using bigger starts. Data that are entirely less than zero can be multiplied by -1 and then re-expressed. When a batch has both positive and negative values, sometimes the positive and negative portions can be re-expressed separately. Other data batches may need special attention beyond the scope of the discussion in this book.

The ladder of powers will prove valuable in a variety of situations throughout this book. The best way to become comfortable with powers is to experiment with the common re-expressions just to see what they do to different data batches. If you can use a computer, it should make such experimentation easy. If not, re-expression is a simple task with a calculator and the letter-value display.

2.5   Re-expression for Symmetry: An Example

To see how re-expression by various powers can help to reshape a batch of data, we now turn to a new set of data. Hinkley (1977) presents data on the amount of precipitation measured during the month of March in 30 consecutive years at Minneapolis/St. Paul. Exhibit 2–5 lists these data and shows a stem-and-leaf display; Exhibit 2–6 gives the letter-value display.

Aside from the isolated value at 4.75, the stem-and-leaf display in Exhibit 2–5 reveals a substantial amount of asymmetry in the batch; the clear

Exhibit 2–5 Thirty Consecutive Values of March Precipitation at Minneapolis/St. Paul

The Data (read across)

0.77	1.74	0.81	1.20	1.95	1.20
0.47	1.43	3.37	2.20	3.00	3.09
1.51	2.10	0.52	1.62	1.31	0.32
0.59	0.81	2.81	1.87	1.18	1.35
4.75	2.48	0.96	1.89	0.90	2.05

Stem-and-Leaf Display
(Unit = .1 Inch of Precipitation in March)

```
                              1   2   represents 1.2

                2       0* |43
                9       0· |7855899
               15       1* |224313
               15       1· |795688
                9       2* |2140
                5       2· |8
                4       3* |300
                        3· |
                        4* |
                1       4· |7
```

Source: Data from D. Hinkley, "On Quick Choice of Power Transformation," *Applied Statistics* 26 (1977):67–69. Reprinted by permission.

Exhibit 2–6 Letter-Value Display for the March Precipitation in Minneapolis/St. Paul Shown in Exhibit 2–5

n = 30

		Lower	Upper	Mid	Spread
M	15.5		1.47	1.47	
H	8	0.90	2.10	1.50	1.20
E	4.5	0.68	2.905	1.79	2.225
D	2.5	0.495	3.23	1.86	2.735
C	1.5	0.395	4.06	2.23	3.665
	1	0.32	4.75	2.535	4.43

upward trend of the midsummaries in Exhibit 2–6 indicates skewness to the right. To move toward symmetry, we should try re-expressions lower on the ladder of powers. Exhibit 2–7 shows the letter-value displays for the square-root, log, and negative-reciprocal re-expressions. Note that the midsummaries for square root, log, and reciprocal are *not* re-expressions of the raw midsummaries. Each midsummary column reports the averages of the letter values of the re-expressed data. Exhibit 2–8 brings together the columns of midsummaries from Exhibits 2–6 and 2–7. As we look for trends down each column of midsummaries in turn, from raw to root to log to reciprocal, we can see the progressively stronger effect of the re-expressions. In the square-root column, the mids still show some upward trend, but the trend is much weaker than in the raw data. The mids in the log column have a stronger *downward* trend, and the mids in the reciprocal column run quite clearly downward. We might try a

Exhibit 2–7   Letter-Value Displays for Minneapolis/St. Paul March Precipitation in Three Expressions (Raw is in Exhibit 2–6.)

Root

		Lower		Upper	Mid	Spread
M	15.5		1.212		1.212	
H	8	0.949		1.449	1.199	0.500
E	4.5	0.822		1.704	1.263	0.881
D	2.5	0.704		1.797	1.250	1.093
C	1.5	0.626		2.008	1.317	1.382
	1	0.566		2.179	1.372	1.614

Log

		Lower		Upper	Mid	Spread
M	15.5		0.167		0.167	
H	8	−0.046		0.322	0.138	0.368
E	4.5	−0.171		0.463	0.146	0.634
D	2.5	−0.306		0.509	0.101	0.815
C	1.5	−0.411		0.602	0.095	1.014
	1	−0.495		0.677	0.091	1.172

Reciprocal

		Lower		Upper	Mid	Spread
M	15.5		−0.681		−0.681	
H	8	−1.111		−0.476	−0.794	0.635
E	4.5	−1.497		−0.345	−0.921	1.152
D	2.5	−2.025		−0.310	−1.168	1.715
C	1.5	−2.626		−0.254	−1.440	2.373
	1	−3.125		−0.211	−1.668	2.914

Exhibit 2–8   Midsummaries for Several Expressions of the Minneapolis/St. Paul March Precipitation

Tag	Raw	Root	Log	Reciprocal
M	1.47	1.212	.1672	−0.681
H	1.50	1.199	.1382	−0.794
E	1.79	1.263	.1458	−0.921
D	1.86	1.250	.1014	−1.168
C	2.23	1.316	.0954	−1.440
1	2.535	1.372	.0909	−1.668

power between root and log, such as the ¼ power, but this batch has only 30 data values—too few for such fine discriminations. If we had to choose among re-expressions listed in Exhibit 2–4, we might select the square root for its simplicity. (Some meteorologists have found the ⅓ power quite desirable.)

* 2.6   Comparing Spreads to the Gaussian Distribution

We have seen how to use the midsummaries to investigate departures from symmetry in a batch. When a batch is roughly symmetric, we can use the spreads to learn still more about its shape. However, the technique we use requires a little more technical detail than we have needed up to now. The basic idea is to compare a symmetric batch to the *Gaussian distribution,* often called the *normal distribution,* on which many traditional statistical techniques are based. Several ways of making this comparison are possible, but this section discusses only one quick and simple method.

Gaussian distribution

normal distribution

Because the Gaussian distribution is symmetric, we begin with a batch of data that is reasonably close to being symmetric, either in its original form or after a re-expression. We then compare the spreads of these data to the corresponding spreads for samples of *n* values from a Gaussian distribution. To keep the calculations simple, we work with the spreads for the *standard Gaussian distribution,* which has mean 0 and standard deviation 1. These spreads are shown in Exhibit 2–9. To obtain spreads for a Gaussian distribution with standard deviation σ, we simply multiply the values in Exhibit 2–9 by σ. Thus, the general value of the Gaussian H-spread is 1.349σ.

standard Gaussian distribution

A simple way to compare the spreads of the data with the Gaussian

Exhibit 2–9   Spreads (at the letter values) for the Standard Gaussian Distribution

Tag	Spread
H	1.349
E	2.301
D	3.068
C	3.726
B	4.308
A	4.836
Z	5.320

spreads is to divide the spread values of the data by the Gaussian spread values:

$$(\text{data H-spread})/1.349,$$

$$(\text{data E-spread})/2.301,$$

$$(\text{data D-spread})/3.068,$$

and so on.

If the data resemble a sample from a Gaussian distribution, then all of these quotients will be nearly the same. In viewing the results, of course, we must remember that the more extreme letter values can be more sensitive to the presence of unusual values in the data.

We can think of each of these calculations as solving for σ. For example, if

$$\text{H-spread} = 1.349\sigma,$$

then $\sigma = \text{H-spread}/1.349$. This is quite different from using the sample standard deviation,

$$s = \sqrt{\frac{1}{n-1}\Sigma(x_i - \bar{x})^2},$$

but the results will be much less affected by stray values. Of course, when the data are not close to Gaussian, 1.349 will not be the correct divisor for the

H-spread. Fortunately, the estimate of σ will not be terribly sensitive to the population shape, at least for the H-spread. As we go to the E-spread or D-spread, sensitivity increases.

A clear trend in the quotients derived from the spreads provides an indication of how the data depart from the Gaussian shape. If the quotients grow, the tails of the batch are heavier than the tails of the Gaussian shape. If the quotients shrink, the tails of the data are lighter.

In Chapter 9 we will see another use of the Gaussian distribution as a standard of comparison.

2.7   Letter Values from the Computer

A letter-value display is simply a table of numbers arranged in columns. The first column contains labels. Columns 2 through 6 contain depths, lower letter values, upper letter values, mids, and spreads, in that order. Computers have little trouble printing such tables. A computer-generated letter-value display usually looks exactly like a neatly typed letter-value display without the ruled lines sometimes used to set off the letter values themselves.

The program must be told which data batch to display. How to tell this to the program depends upon the particular implementation of the program. All decisions are made automatically, so no further information is needed.

† 2.8   Algorithms

The FORTRAN and BASIC programs for letter values work in slightly different ways, illustrating two alternative organizations of the tasks involved. The FORTRAN program finds all the letter values first and places them and their depths in arrays for subsequent printing. This has the advantage of making the letter values available for other computations. The BASIC version prints the letter values as it finds them and uses no additional storage. The BASIC program also attempts to position the columns of the display in order to make the best use of the available page area.

It is difficult for portable programs to control the number of decimal places printed and to align the decimal points of the numbers in each column.

Implementers of the FORTRAN version may want to use run-time formats to avoid the possibility of a number's overflowing the formatted size allowed here. Implementers of the BASIC version who have a PRINT USING statement available in their BASIC may wish to use it to format the columns.

FORTRAN

The FORTRAN programs for finding letter values and displaying them consist of two subroutines: LVALS and LVPRNT. LVALS accepts the data in a vector and returns a vector of depths and an array of pairs of letter values. It is used through the statement

        CALL LVALS(Y, N, D, YLV, NLV, SORTY, ERR)

where the arguments are as follows:

Y( )	is the N-long vector of data values;
N	is the number of data values;
D(15)	is the vector of depths;
YLV(15,2)	is the array of letter values [YLV(1,1) and YLV(1,2) both contain the median, and the remaining pairs of letter values are in order from the hinges out to the extremes, with the lower letter value first];
NLV	returns the number of pairs of letter values;
SORTY( )	is the N-long workspace for sorting Y( );
ERR	is the error flag, whose values are

	0	normal
	21	$N < 2$ or $N > 24576$—too few or too many data values
	22	$NLV < 3$ or $NLV > 15$
	23	page width < 64 print positions—too narrow for letter-value display.

The subroutine LVPRNT uses the information on depths and letter values to print the letter-value display in essentially the format shown in Exhibits 2–3 and 2–6. The calling statement is

        CALL LVPRNT(NLV, D, YLV, ERR)

where the arguments are as described above.

BASIC

The BASIC program requires only the defined functions and the SORT from Y( )
to W( ) subroutines. It leaves X( ) and Y( ) unchanged.

2.9   Sorting

sorting

The process of putting a set of numbers or other elements, such as names, into
order is known as *sorting*. Because an ordered batch makes it easy to pick out
the letter values, as well as to detect potentially stray values at either end,
sorting is an important operation in exploratory data analysis. This section
discusses the reasons for including certain sorting programs in this book; it also
provides selected references so that interested readers can pursue the subject
of sorting further.

Computer scientists have devoted considerable imagination and energy
to designing and analyzing algorithms for sorting. Their analyses tell us,
among other things, how much time a given sorting algorithm requires to
process a batch of n numbers when n is large. For some algorithms this time is
proportional to n^2. This is easy to understand if we imagine making $n - 1$
comparisons to pick out the smallest number, $n - 2$ comparisons to find the
next smallest, and so on. The total number of comparisons is $(n - 1) +
(n - 2) + \ldots + 1 = n(n - 1)/2$, which resembles $n^2/2$ when n is large.

However, it is possible to sort much more efficiently than in time
proportional to n^2. Fast sorting algorithms require time proportional to
$n \log(n)$, and the difference between $n \log(n)$ and n^2 becomes greater as n
increases. If we want only a few values at selected positions in the ordered
batch, we can even obtain these values without sorting the batch completely.
Such a "partial sorting" algorithm could, for example, deliver the median in
time proportional to n.

The sorting algorithms used in the programs in this book are not the
most elegant algorithms available, but they are among the simplest to
program. Their simplicity makes them easier to read and understand, and they
take up much less space than do the faster methods—both are an advantage on
small computers. Also, users of these programs will often be concerned only
with situations in which n is small—for example, n less than 50—and the
greater effort that the fast algorithms put into bookkeeping may not be
worthwhile. Sorting programs for a variety of applications are available in

most computing environments; it may be easier to use one of these, provided that it can be called in the same way, than to adopt the simple programs in this book.

Two references provide useful additional information about sorting algorithms. In a careful tutorial paper Martin (1971) discusses a considerable variety of sorting techniques and the circumstances under which they are appropriate. Aho, Hopcroft, and Ullman (1974) use several important and interesting sorting techniques to illustrate the analysis of sorting algorithms and include a careful discussion of partial sorting.

References

Aho, Alfred V., John E. Hopcroft, and Jeffrey D. Ullman. 1974. *The Design and Analysis of Computer Algorithms*. Reading, Mass.: Addison-Wesley.

Hinkley, David V. 1977. "On Quick Choice of Power Transformation." *Applied Statistics* 26:67–69.

Martin, William A. 1971. "Sorting." *Computing Surveys* 3:147–174.

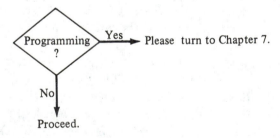

```
5000 REM    LETTER-VALUE DISPLAY
5010 REM    PRINT A LETTER-VALUE DISPLAY FOR THE DATA IN Y() OF LENGTH N.
5020 REM    VERSION V1=1 PRINTS 7-NUMBER SUMMARY ONLY.
5030 REM
5040 REM    SORT Y() INTO W()

5050 GOSUB 3300

5060 REM SET UP TABSTOPS FOR COLUMNS

5070 LET T9 = FNI((M9 - M0 - 1) / 5)
5080 LET T1 = M0 + 2
5090 LET T2 = T1 + T9
5100 LET T3 = T2 + T9
5110 LET T4 = T3 + T9
5120 LET T5 = T4 + T9

5130 REM    SET UP TRUNCATION DECIMAL PLACE

5140 LET T8 = ABS( FNI( FNL(W(1)))) + 4
5150 IF T8 < T9 THEN 5170
5160 LET T8 = T9 - 1

5170 REM    PRINT HEADING

5180 PRINT
5190 PRINT TAB(T1);"DEPTH"; TAB(T2);"LOW"; TAB(T3 + 1);"HIGH";
5200 PRINT TAB(T4 + 2);"MID"; TAB(T5);"SPREAD"
5210 PRINT

5220 REM    MEDIAN LINE IS SPECIAL

5230 LET K = FNI(N + 1) / 2
5240 LET W1 = FNT( FNM(K))
5250 PRINT TAB(M0);"M"; TAB(T1);K;
5255 PRINT TAB( FNI((T2 + T3) / 2 + 2 - LEN( STR$(W1)) / 2));W1;
                                                        TAB(T4);W1

5260 REM    INITIALIZE LABELS; L$ TO PRINT, L TO COUNT IN ASCII
5270 REM    NOTE THAT THIS CODE IS ASCII-DEPENDENT, ALTHOUGH MODIFICATION
5280 REM    TO OTHER CHARACTER CODES SHOULD BE SIMPLE.

5290 LET L$ = "H"
5300 LET L = ASC("E")
```

```
5310 REM   NOW LOOP TO PRINT LETTER VALUES. K COUNTS DEPTHS

5320 LET K = FNI(K + 1) / 2
5330 LET W1 = FNM(K)
5340 LET W2 = FNM(N - K + 1)
5350 PRINT TAB(M0);L$; TAB(T1);K; TAB(T2); FNT(W1); TAB(T3); FNT(W2);

5360 PRINT TAB(T4); FNT((W1 + W2) / 2); TAB(T5); FNT(W2 - W1)
5370 LET L$ = CHR$(L)
5380 LET L = L - 1
5390 IF L >= ASC("A") THEN 5410
5400 LET L = ASC("Z")
5410 IF V1 > 1 THEN 5440

5420 REM BRIEF VERSION STOPS AT 7-NUMBER SUMMARY--DID WE JUST DO E'S?

5430 IF L$ = "D" THEN 5460

5440 REM   LOOP IF THERES MORE TO DO

5450 IF K > 2 THEN 5310

5460 REM   PRINT EXTREMES AND EXIT

5470 PRINT TAB(T1);"1"; TAB(T2); FNT(W(1)); TAB(T3); FNT(W(N));
5480 PRINT TAB(T4); FNT((W(1) + W(N)) / 2); TAB(T5); FNT(W(N) - W(1))
5490 PRINT
5500 RETURN
```

```
      SUBROUTINE LVALS(Y, N, D, YLV, NLV, SORTY, ERR)
C
      INTEGER N, NLV, ERR
      REAL Y(N), D(15), YLV(15,2), SORTY(N)
C
C FOR THE BATCH OF VALUES IN Y, FIND THE SELECTED QUANTILES KNOWN
C AS THE LETTER VALUES.  UPON EXIT,  YLV  CONTAINS
C THE LETTER VALUES,  D  CONTAINS THE CORRESPONDING
C DEPTHS, AND  NLV  IS THE NUMBER OF PAIRS OF
C LETTER VALUES.  SPECIFICALLY,  YLV(1,1)  AND
C YLV(1,2)  ARE BOTH SET EQUAL TO THE MEDIAN, WHOSE DEPTH,
C D(1),  IS (N+1)/2 .  THE REST OF THE LETTER VALUES
C COME IN PAIRS AND ARE STORED IN  YLV  IN ORDER FROM THE
C HINGES OUT TO THE EXTREMES.  THUS  YLV(2,1) AND
C YLV(2,2)  ARE THE LOWER HINGE AND THE UPPER HINGE,
C RESPECTIVELY, AND  YLV(NLV,1) AND  YLV(NLV,2)  ARE THE
C LOWER EXTREME (MINIMUM) AND UPPER EXTREME (MAXI-
C MUM), RESPECTIVELY.
C
C LOCAL VARIABLES
C
      INTEGER I, J, K, PT1, PT2
C
      IF((N .GT. 3) .AND. (N .LE. 24576)) GO TO 10
        NLV = 0
        ERR = 21
        GO TO 999
C
C SORT Y INTO SORTY
C
   10 DO 15 I = 1,N
        SORTY(I) = Y(I)
   15 CONTINUE
      CALL SORT(SORTY, N, ERR)
      IF(ERR .NE. 0) GO TO 999
C
C HANDLE MEDIAN SEPARATELY BECAUSE IT IS NOT A PAIR
C OF LETTER VALUES.
C
      D(1) = FLOAT(N + 1) / 2.0
      J = (N / 2) + 1
      PT2 = N + 1 - J
      YLV(1,1) = (SORTY(J) + SORTY(PT2)) / 2.0
      YLV(1,2) = YLV(1,1)
C
      K = N
      I = 2
C
   20 K = (K + 1) / 2
      J = (K / 2) + 1
      D(I) = FLOAT(K + 1) / 2.0
```

61

```
C
      PT2 = K + 1 - J
      YLV(I,1) = (SORTY(J) + SORTY(PT2)) / 2.0
      PT1 = N - K + J
      PT2 = N + 1 - J
      YLV(I,2) = (SORTY(PT1) + SORTY(PT2)) / 2.0
C
      I = I + 1
      IF(D(I-1) .GT. 2.0) GO TO 20
C
      NLV = I
      D(I) = 1.0
      YLV(I,1) = SORTY(1)
      YLV(I,2) = SORTY(N)
C
  999 RETURN
      END



      SUBROUTINE LVPRNT(NLV, D, YLV, ERR)
C
      INTEGER NLV, ERR
      REAL D(15), YLV(15,2)
C
C  PRINT A LETTER-VALUE DISPLAY.
C  THE  NLV  PAIRS OF LETTER VALUES ARE IN  YLV
C  -- YLV(I,1)  IS THE LOWER LETTER VALUE IN
C  THE PAIR AND  YLV(I,2)  IS THE UPPER LETTER
C  VALUE, WITH THE EXCEPTION THAT  YLV(1,1)
C  AND  YLV(1,2)  ARE BOTH EQUAL TO THE MEDIAN.
C  THE VECTOR  D  CONTAINS THE CORRESPONDING
C  DEPTHS.
C
      COMMON /CHRBUF/ P, PMAX, PMIN, OUTPTR, MAXPTR, OUNIT
      INTEGER P(130), PMAX, PMIN, OUTPTR, MAXPTR, OUNIT
C
C  LOCAL VARIABLES
C
      INTEGER I, N, TAGS(14)
      REAL MID, SPR
C
      DATA TAGS( 1), TAGS( 2), TAGS( 3), TAGS( 4) /1HM, 1HH, 1HE, 1HD/
      DATA TAGS( 5), TAGS( 6), TAGS( 7), TAGS( 8) /1HC, 1HB, 1HA, 1HZ/
      DATA TAGS( 9), TAGS(10), TAGS(11), TAGS(12) /1HY, 1HX, 1HW, 1HV/
      DATA TAGS(13), TAGS(14)                     /1HU, 1HT/
C
      IF((NLV .GE. 3) .AND. (NLV .LE. 15)) GO TO 10
        ERR = 22
        GO TO 999
   10 IF(PMAX .GE. 64) GO TO 20
        ERR = 23
        GO TO 999
   20 WRITE(OUNIT, 1001)
 1001 FORMAT(5X,5HDEPTH,7X,5HLOWER,8X,5HUPPER,11X,
     1  3HMID,8X,6HSPREAD)
```

```
C
C   RECOVER  N  FROM  D(1) , THE DEPTH OF THE MEDIAN.
C
      N = INT(2.0 * D(1)) - 1
      WRITE(OUNIT, 1002) N
 1002 FORMAT(1X,2HN=,I5)
C
C   WRITE LINE CONTAINING MEDIAN (AND FIRST MID).
C
      WRITE(OUNIT, 1003) D(1), YLV(1,1), YLV(1,1)
 1003 FORMAT(1X,1HM,1X,F7.1,8X,F10.3,13X,F10.3)
C
      N = NLV - 1
      DO 30 I = 2, N
        MID = (YLV(I,1) + YLV(I,2)) / 2.0
        SPR = YLV(I,2) - YLV(I,1)
        WRITE(OUNIT, 1004) TAGS(I), D(I), YLV(I,1),
    1   YLV(I,2), MID, SPR
 1004   FORMAT(1X,A1,1X,F7.1,3X,F10.3,3X,F10.3,5X,F10.3,3X,F10.3)
   30 CONTINUE
      MID = (YLV(NLV,1) + YLV(NLV,2)) / 2.0
      SPR = YLV(NLV,2) - YLV(NLV,1)
      WRITE(OUNIT, 1005) YLV(NLV,1), YLV(NLV,2), MID, SPR
 1005 FORMAT(7X,1H1,5X,F10.3,3X,F10.3,5X,F10.3,3X,F10.3/)
C
  999 RETURN
      END
```

Chapter 3 _____

Boxplots

In Chapter 1 we saw that stem-and-leaf displays provide a flexible and effective way to view a batch of data as a whole. In Chapter 2 we considered a numerical summary of a batch using a few values at selected depths. Frequently, we can make good use of something between these two extremes in the form of a picture or graphical summary. We want to represent the data values graphically, but we do not want to see all the detail. This is just the task for which boxplots were invented.

3.1 Basic Purposes

Most batches of data pile up in the middle and spread out toward the ends. To summarize the behavior of a batch, we need a clear picture of where the middle lies, roughly how spread out the middle is, and just how the tails relate to it. Since the middle is generally better defined than the tails of the data, we need to see less detail at the middle—we want to focus more of our attention on possible strays at the ends because these often give clues to unexpected behavior. To some extent, a letter-value display focuses numerically on the ends because the depths of the letter values are selected to give increasing detail toward the extremes. We could represent the letter-value display graphically, but we would find ourselves paying too much attention to end values that fit in well with the rest of the batch. What we need is a rule for showing only values that are unusually extreme and hence are likely to be strays. When we have several related batches, we can learn more about symmetry and strays by comparing those batches. When we have only one batch, we must depend on the middle to help us identify strays at the ends.

3.2   The Skeletal Boxplot

5-number summary

If we wanted to turn a letter-value display into a graph, we could begin with the simplest letter-value display, the *5-number summary,* which gives median, hinges, and extremes. For the areas of New Jersey counties from Exhibits 2–1 and 2–3, the 5-number summary is Exhibit 3–1. Exhibit 3–2 presents these letter values graphically. It is an example of a skeletal box-and-whiskers plot, or skeletal boxplot for short. It shows the middle of the batch, from hinge to hinge, as a box with a line through it at the median, and it runs a solid "whisker" out from each hinge to the corresponding extreme. With one glance the eye can easily form impressions of overall level, amount of spread, and symmetry. Thus, in Exhibit 3–2 we see that the median is around 300, the H-spread is around 250, and the range of the data is roughly 800. These data depart somewhat from symmetry—the median lies below the middle of the box, and the upper whisker is nearly twice as long as the lower one.

Exhibit 3–1   Five-Number Summary for Areas of New Jersey Counties Shown in Exhibit 2–1

$n = 21$

		Lower		Upper	Mid	Spread
M	11		329		329	
H	6	228		476	352	248
	1	47		819	433	772

Exhibit 3–2   Box-and-Whiskers Plot for Areas of New Jersey Counties

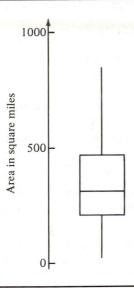

3.3   Outliers

outliers Some data batches include *outliers,* values so low or so high that they seem to stand apart from the rest of the batch. Some outliers may be caused by measuring, recording, or copying errors or by errors in entering the data into the computer. When such errors occur, we will want to detect and correct them, if possible. If we cannot correct them (but believe they are in error), we will probably want to exclude the erroneous values from further analysis.

Not all outliers are erroneous. Some may merely reflect unusual circumstances or outcomes; so having these outliers called to our attention can help us to uncover valuable information. Whatever their source, outliers demand and deserve special attention. Sometimes we will try to identify and display them; other times we will try to insulate our analyses and plots from their effects. In succeeding chapters we will continue to examine outliers in one way or another.

To deal with outliers routinely, we need a rule of thumb that the computer can use to identify them. For this we use the hinges and their *inner fences*   difference, the H-spread. We define the *inner fences* as

$$\text{lower hinge} - (1.5 \times \text{H-spread})$$

$$\text{upper hinge} + (1.5 \times \text{H-spread})$$

outer fences   and the *outer fences* as

$$\text{lower hinge} - (3 \times \text{H-spread})$$

$$\text{upper hinge} + (3 \times \text{H-spread}).$$

outside
far outside

adjacent
value

Any data value beyond either inner fence we term *outside,* and any data value beyond either outer fence we call *far outside*. The outermost data value on each end that is still not beyond the corresponding inner fence is known as an *adjacent value*.

For the New Jersey counties example, we have seen (in Exhibit 3–1) that the hinges are at 228 and 476 square miles, so the H-spread is $476 - 228 = 248$. Thus, the inner fences are at

$$228 - 1.5 \times 248 = 228 - 372 = -144$$

$$476 + 1.5 \times 248 = 476 + 372 = 848$$

and the outer fences are at

$$228 - 3 \times 248 = 228 - 744 = -516$$

$$476 + 3 \times 248 = 476 + 744 = 1220.$$

Because neither of the extreme values is "outside," the adjacent values are the extremes, 47 and 819.

By contrast, the precipitation pH data, which appear in Exhibit 1–1, have three outside values. From the letter-value display in Exhibit 3–3, we see that the hinges are 4.31 and 4.82. Thus, the inner fences are 3.545 and 5.585, and the outer fences are 2.78 and 6.35. The three data values 5.62, 5.67, and 5.78 are outside, and thus, by this rule of thumb, outlying. The adjacent values are 4.12 and 5.51.

We can also use this rule for identifying outliers in the stem-and-leaf display. If outside values appear on the special LO and HI stems, the LO and

Exhibit 3-3 Letter-Value Display for pH Values of Precipitation in Allegheny County, Pennsylvania

		Lower		Upper	Mid	Spread
n = 26						
M	13.5		4.54		4.54	
H	7	4.31		4.82	4.565	0.51
E	4	4.26		5.51	4.885	1.25
D	2.5	4.19		5.645	4.92	1.455
C	1.5	4.12		5.725	4.92	1.605
	1	4.12		5.78	4.95	1.66

Note: Data from Exhibit 1–1.

HI stems serve the dual purpose of highlighting the outliers for special attention and preserving a useful choice of scale. Otherwise, we might have many empty stems between an outlier and the body of the data (see Appendix A for more details). We can modify the skeletal boxplot to include information about outliers, as we see in the next section.

3.4 Making a Boxplot

boxplot

We begin a *boxplot* in the same way as we begin a skeletal boxplot: We use solid lines to mark off a box from hinge to hinge and show the median as a solid line across it. Next, we run a dashed whisker out from each hinge *to the corresponding adjacent value* instead of to the extreme, as in the skeletal boxplot. Then we show each outside value individually and, if each data value has an identity, as often happens, label it clearly. Finally, we show each far outside value individually and label it quite prominently—for example, with a tag in capital letters. When it is informative and will not cause clutter, we may also label the adjacent values. Because the fences are not necessarily data values, we do not mark them; they simply serve to define outside and far outside values. The original name for the boxplot is "schematic plot." But, the convenience of a short, suggestive name has led most people who use it to refer to the display as a boxplot, and that term is used in this book.

Exhibit 3–4   Boxplot for the Precipitation pH Data

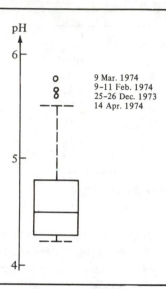

For the precipitation pH data, we have done all the necessary calculations in the previous section, and Exhibit 3–4 shows the boxplot. The three outside values are clearly evident, so we look more closely at them. A careful look at the data—see, for example, Exhibit 1–3—indicates that although the fourth largest value (5.51) has not been identified as an outlier by the rule of thumb, it resembles the three outside values more than it does the rest of the batch. (We might have suspected this from the long upper whisker.) This example highlights the important lesson that the rule of thumb for outliers is no more than a convenient guideline and is no substitute for good judgment. We would probably choose to treat all four of these values as potential outliers.

In the precipitation pH data, there is little reason to suspect errors in the data, so we look up the dates of the precipitation samples in Exhibit 1–1 and use them as labels on the boxplot. Three of the four dates are holidays: 14 Apr. 1974 was Easter, 25–26 Dec. 1973 was Christmas, and 12 Feb. 1974 was Lincoln's birthday and fell on a Monday. Is there something unusual about holiday weekends? Recall that the original study was motivated by the suspicion that air pollution contributes to making rain more acidic. The outliers are the least acidic observations. The other outside value, 9 Mar. 1974, does not correspond to a holiday; but if more data were available, we would now want to try separating holiday and non-holiday periods.

3.5   Boxplots from the Computer

The most obvious difference between boxplots produced by the programs in this chapter and boxplots drawn by hand is that the computer-produced plots are drawn across the page. The horizontal format is quicker to print than is a vertical plot on most computer terminals and makes it easy to produce a number of boxplots side by side for comparing batches.

Because most computer terminals cannot draw pictures, we must construct boxes out of the normal printing characters. BASIC and FORTRAN, the two computer languages used here, have different character sets. BASIC uses the standard ASCII character set found on most terminals; the standard FORTRAN character set is much more limited (see Appendix C for details). Thus, what a computer-produced boxplot looks like on your computer may depend on which set of programs—that is, which language—is used and on decisions made when the programs are implemented.

The BASIC version of a computer-produced boxplot looks like this:

```
             ----------------------
    --------[         +           ]--------------------------    *  *  *
             ----------------------
```

The box is formed with two square brackets and two lines of minus signs. The location of the median is marked with a +. The whiskers are dashed lines as in handmade boxplots, and outliers are marked with an asterisk (outside) or a capital O (far outside). A simpler form, the 1-line boxplot, omits the dashed lines that complete the box:

```
    --------[         +           ]--------------------------    *  *  *
```

The FORTRAN version looks like this:

```
             ----------------------
    --------I         +           I--------------------------    *  *  *
             ----------------------
```

The only difference between the two versions is the use of the letter I in place of the square brackets.

3.6   Comparing Batches

Often we may want to place several boxplots side by side to compare several batches. For example, Exhibit 3–5 gives the percentages of individual tax

Exhibit 3–5   Percentages of Individual Tax Returns Audited by the IRS in the States of the United States in Fiscal Year 1974

North Atlantic	*Percentage*	*Midwest*	*Percentage*
New York	3.0	South Dakota	1.5
Maine	2.1	North Dakota	1.8
Massachusetts	1.6	Illinois	2.0
Vermont	2.1	Iowa	1.3
Connecticut	1.8	Wisconsin	1.7
New Hampshire	2.2	Nebraska	2.3
Rhode Island	1.8	Missouri	2.1
Mid-Atlantic		Minnesota	1.4
Maryland & D.C.	2.1	*Southwest*	
New Jersey	2.1	New Mexico	2.1
Pennsylvania	1.6	Wyoming	1.8
Virginia	1.9	Colorado	1.9
Delaware	2.2	Texas	3.1
Southeast		Arkansas	2.2
Georgia	2.3	Louisiana	2.6
Alabama	2.3	Oklahoma	2.3
South Carolina	2.3	Kansas	2.5
North Carolina	2.7	*West*	
Mississippi	2.8	Alaska	2.7
Florida	2.7	Idaho	2.0
Tennessee	1.7	Montana	2.7
Central		Hawaii	1.9
Ohio	1.4	California	2.5
Michigan	2.0	Arizona	1.9
Indiana	1.2	Oregon	1.5
Kentucky	1.4	Nevada	3.4
West Virginia	1.3	Utah	2.2
		Washington	2.0

Source: Data from *1976 Tax Guide for College Teachers* (Washington, D.C.: Academic Information Service, Inc., 1975) pp. 195–197. Reprinted by permission.

returns audited by the Internal Revenue Service (IRS) in the states of the United States in fiscal year 1974. To look into possible regional differences in the auditing rate, we can begin with the boxplots shown in Exhibit 3–6. We note that auditing rates seem comparatively low for the Central states, except for one far outside state, which, we can see from Exhibit 3–5, is Michigan. Conversely, the Southeast seems to have relatively high audit rates for the eastern United States, except for the low outside value for Tennessee. Western

Exhibit 3–6   Side-by-Side Boxplots of the IRS Audit Rates of Exhibit 3–5

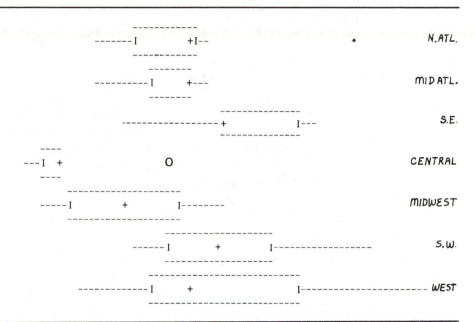

states include the highest auditing rate, Nevada at 3.4%, but are quite spread out. We note also that three batches have medians that coincide with a hinge, so that the + marking the median overprints the hinge marker. This is due in part to the small number of states in some regions and in part to several states having the same audit rate.

* 3.7   More Refined Comparisons: Notched Boxplots

When we use boxplots to compare batches, we are tempted to note batches that are "significantly" different from each other or from some standard batch. Our eyes tend to look for non-overlapping central boxes; but unfortunately the hinges, which determine the extent of the box, are inappropriate guides to significance. McGill, Tukey, and Larsen (1978) have shown one way to use regions of overlap or non-overlap of special intervals around each *notch*   median of a boxplot. They mark the ends of these intervals by putting a ***notch*** in the side of the central box. Two groups whose notched intervals do not overlap can be said to be significantly different at roughly the 5% level. (This is an *individual* 5% level—that is, no allowance is made for the number of comparisons considered.)

    The notches in these plots are placed symmetrically around the median, falling at

$$\text{median} \pm 1.58 \times (\text{H-spr})/\sqrt{n}.$$

The multiplying factor, 1.58, combines contributions from three different sources: the relationship between the H-spread and the (population) standard deviation, the variability of the sample median, and the factor used in setting confidence limits. The details underlying the choice of 1.58 are given in Section 3.12 at the end of this chapter.

    Computer-produced boxplots indicate notches on the main line of the display. A notched boxplot in BASIC looks like this:

```
    --------[ > + < ]--------    *    *    *      O      O
```

In the FORTRAN programs, the notches are marked with parentheses:

```
                  -------------------------
    -----------------I(      +      )    I-------------------
                  -------------------------
```

    Exhibit 3–7 shows the audit data of Exhibit 3–5 with notches added. We note that in some regions, and especially when the median is near a hinge, one of the notches actually falls outside the box. Now we can see, for example, that, although we might have been tempted to declare the median audit rates for the Mid-Atlantic and Southeast regions significantly different, we cannot be confident of this difference at the 5% level.

3.8   Using the Programs

The boxplot programs are quite automatic. They produce a display for whatever data batch is specified. (Again, how you specify a data batch to your computer depends upon how the programs have been implemented.) The options offered by the boxplot programs are the choice of a 1-line or 3-line display and the inclusion or exclusion of notches. The 3-line version looks more like the hand-drawn boxplot and may be preferred for single batches. However, because multiple 3-line boxplots can become cluttered and may take up too many lines on a CRT screen,* we often use the 1-line display to compare more than 3 or 4 groups. One-line *notched* boxplots can be particularly useful for comparing batches.

---

*A Cathode Ray Tube (CRT) is like a television screen and is used in many computer terminals to display output. Often it can display only 20 lines or so at a time.

Exhibit 3–7   Multiple Notched Boxplots to Compare IRS Audit Rates of Exhibit 3–5

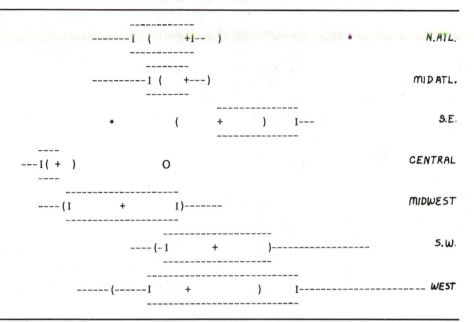

Multiple boxplots require additional information—namely, the identity of the group to which each data value belongs. The programs distinguish groups by using consecutive identifying integers, starting with 1. Because data values are not always arranged according to groups, we must provide this information by telling the computer which group the first data value belongs to, and so on. One possible source of group identity is the column number or the row number of data values in a table. We examine tables of data in Chapter 7.

† 3.9 Algorithms

The boxplot programs must place the pieces of the boxplot display in the correct printing positions (see Appendix A for a discussion of computer graphics). In addition, the programs must take care that if two characters making up the display fall at the same printing position, the one actually printed will convey as much information about the plot as possible. The programs accomplish this by first constructing each line of the boxplot display in an array and then printing the contents of the array.

    Characters are positioned on the output line according to the plot scale. The logical width of one character position, called the ***nice position width***, *NPW,* is found by using the utility plot-scaling routines (see Appendix A). The number of the printing position that corresponds to the data value, *y*, can then be found as

nice position width

$$[(y - \min(y))/NPW] + 1$$

where $\min(y)$ is the minimum data value and [ ] indicates the integer part.

    The programs ensure correct priority of plot symbols by placing them in the output array in a specified order, allowing later entries to replace earlier ones if they fall at the same character position. The correct placement order—and, hence, the order from least important to most important—is: whisker hyphens (-); outside values (*); far outside values (O); hinges ([ ] or I); notches, if any (> < or ( )); and median (+). It is usually easy to read even severely distorted boxplots generated in this order. Thus,

$$---[+----$$

is a boxplot in which the H-spread is small and the median is offset to the high end and thus occupies the same position as the upper hinge. In a very extreme case,

                       * + *   OO

is a display in which most of the data clusters very near the median and there are a few very extreme outliers. Exhibit 3–7 includes several boxes in which overprinting is evident. In each of these, the careful choice of symbol hierarchy has preserved the full information in the plot. Multiple boxplots require a single scale that is usually chosen to cover the range of the entire combined data set.

FORTRAN

The FORTRAN programs for creating and displaying boxplots consist of three subroutines, BOXES, BOXP, and BOXTOP, and the function PLTPOS. The display of one or several boxplots is initiated by the statement

CALL BOXES(Y, N, GSUB, NG, LINE3, NOTCH, SORTY, ERR)

where the parameters have the following meanings:

Y( )	is an array of N data values;
N	is the number of data values;
GSUB( )	holds the N group identifiers, integers from 1 to NG, if more than one boxplot is to be produced;
NG	is the number of groups and thus the number of boxplots to be displayed;
LINE3	is a logical flag, set .TRUE. for a 3-line plot, or set .FALSE. for a 1-line plot;
NOTCH	is a logical flag, set .TRUE. for notched boxplots;
SORTY( )	is an N-long work array in which to sort Y( ) or groups;
ERR	is the error flag, whose values are
	0 normal
	31 N < 2—too few data values to make a boxplot.

BOXES determines the plot scaling (see Appendix A) and calls BOXP for each boxplot. BOXP, in turn, calls BOXTOP to produce the top and bottom of any 3-line boxplot and uses the function PLTPOS in placing symbols in the output array.

BASIC

The BASIC programs for boxplots accept N data values in Y( ). The style of boxplot is determined by the version number, V1, where

V1 = 1	1-line boxplot,
V1 = 2	1-line notched boxplot,
V1 = 3	3-line boxplot,
V1 = 4	3-line notched boxplot.

If V1 < 0, the program asks for data bounds and uses only the data values falling between these bounds, and the plot style corresponding to |V1|.

The program also checks a secondary version flag, V2. If V2 ≠ 0, the program looks in the subscript array C( ) for group identifiers and prints a boxplot for each group. Group identifiers may be any unique numbers;

sequential integers are simplest. Multiple boxplots use a single global scale and are printed in group-number order. Each group is labeled with its group number, if that label is less than 5 characters long. Boxplots are scaled to fit between the margins, M0 and M9.

The BASIC program does not change X( ) or R( ).

† 3.10   Implementation Details

The boxplot programs depend on the available character set more heavily than do any of the other computer programs in this book. FORTRAN programmers are likely to have available a larger set of characters than are in the FORTRAN standard. They may wish to substitute non-FORTRAN characters when these are available.

The variable that identifies the groups for a multiple boxplot should be implemented as a data vector if at all possible. Note that we have also used data vectors to hold row and column subscripts for tables in Section 7.3.

† 3.11   Further Refinements in Display

Many readers may have available a device that enables their computer to draw displays made up of lines. Boxplots are very well suited to many of these computer graphics devices because boxplots consist almost entirely of vertical and horizontal lines. The same principles used to determine the scale of the plots in the programs provided in this chapter can be used for such displays.

In the paper mentioned in Section 3.7, McGill, Tukey, and Larsen suggest making the width of a boxplot (the fatness of the box) proportional to $\sqrt{n}$, the square root of the batch count. While we could approximate a variable-width boxplot with a "variable-line" boxplot, printer plotting does not provide sufficient precision to justify the trouble. Readers with access to more sophisticated graphics devices that are capable of drawing lines may wish to experiment with this idea.

* 3.12   Details of the Notched Boxplot

The notches in a notched boxplot define a confidence interval around the median that has been adjusted to make it appropriate for comparisons of two boxes. If the intervals of two boxes do not overlap, we can be confident at about the 95% level that the two population medians are different. The notches are placed at

$$\text{median} \pm 1.58 \times (\text{H-spr})/\sqrt{n}.$$

The factor 1.58 combines contributions from three different sources as described in Section 3.7. We now consider the details of these contributions.

First, from the discussion in Section 2.6, we recall that H-spr/1.349 provides a rough estimate of the standard deviation, σ, especially in large samples from a Gaussian distribution.

Another large-sample result from the Gaussian distribution is that the variance of the sample median is $\pi/2$ times the variance of the sample mean. Although this result is strictly true only for large samples from the Gaussian distribution, it turns out to be a surprisingly good estimate for a wide variety of distributions.

Finally, we recall that the usual 95% confidence interval for the mean of a Gaussian distribution with known variance is $\bar{x} \pm 1.96\,\sigma_{\bar{x}}$, where $\sigma_{\bar{x}} = \sigma/\sqrt{n}$. In comparing batches we must face the separate variability of each batch.

If we compare two equally variable batches, we look at

$$\frac{\bar{x}_2 - \bar{x}_1}{\sqrt{\text{var}(\bar{x}_2) + \text{var}(\bar{x}_1)}} = \frac{\bar{x}_2 - \bar{x}_1}{\sqrt{2}\,\sigma_{\bar{x}}} \tag{1}$$

which is a z-score and should thus be compared to ± 1.96. Equivalently, we could compare

$$\frac{|\bar{x}_2 - \bar{x}_1|}{\sqrt{2}\,\sigma_{\bar{x}}} - 1.96 = \frac{|\bar{x}_2 - \bar{x}_1| - 1.96\sqrt{2}\,\sigma_{\bar{x}}}{\sqrt{2}\,\sigma_{\bar{x}}} \tag{2}$$

to zero or simply compare the numerator,

$$|\bar{x}_2 - \bar{x}_1| - 1.96\sqrt{2}\,\sigma_{\bar{x}} \tag{3}$$

to zero; that is, if (3) is greater than zero, we declare the means to be significantly different. To represent this calculation as a comparison between two possibly overlapping confidence intervals for the two means, we split the constant equally between the two intervals and (assuming that $\bar{x}_1 < \bar{x}_2$) compare the upper bound of the lower interval,

$$\bar{x}_1 + \frac{1.96 \sqrt{2} \, \sigma_{\bar{x}}}{2} = \bar{x}_1 + \frac{1.96}{\sqrt{2}} \, \sigma_{\bar{x}} \qquad (4)$$

to the lower bound of the upper interval,

$$\bar{x}_2 - \frac{1.96}{\sqrt{2}} \, \sigma_{\bar{x}}. \qquad (5)$$

This comparison is equivalent to just rewriting (3) as

$$(\bar{x}_2 - \frac{1.96}{\sqrt{2}} \, \sigma_{\bar{x}}) - (\bar{x}_1 + \frac{1.96}{\sqrt{2}} \, \sigma_{\bar{x}}), \qquad (6)$$

which we again compare to zero. Thus, the appropriate constant for constructing confidence intervals for the special case of comparing two equally variable means is not 1.96, but $1.96/\sqrt{2} = 1.39$.

By contrast, if the variances of the two batches were very different—for example, if $\sigma_{\bar{x}_1}^2$ were tiny and $\sigma_{\bar{x}_2}^2$ enormous—we would still compare the means by using

$$\frac{\bar{x}_2 - \bar{x}_1}{\sqrt{\text{var}(\bar{x}_2) + \text{var}(\bar{x}_1)}}. \qquad (7)$$

But now $\text{var}(\bar{x}_2)$ dominates the denominator; so this expression is almost equal to

$$\frac{\bar{x}_2 - \bar{x}_1}{\sigma_{\bar{x}_2}}. \qquad (8)$$

As in equation (1), we compare this to 1.96. The expression corresponding to (3) is

$$\bar{x}_2 - \bar{x}_1 - 1.96\sigma_{\bar{x}_2}, \qquad (9)$$

which we would compare to zero as we did for (3).

In setting intervals to represent this situation, we are led to allocate the variability in $\sigma_{\bar{x}_2}$ to $\bar{x}_2$ and to put back in the negligible variability of $\bar{x}_1$

measured by $\sigma_{\bar{x}_1}$. We thus use

$$\bar{x}_2 \pm 1.96\sigma_{\bar{x}_2}$$

and

$$\bar{x}_1 \pm 1.96\sigma_{\bar{x}_1}.$$

The two extreme situations just described lead to using 1.39 and 1.96 as approximate multiplying constants for these intervals. A reasonable compromise for the general case is the average of the two constants:

$$(1.96 + 1.39)/2 = 1.7.$$

Assembling the three factors—the estimate of σ from the H-spread, the standard deviation of the median relative to the mean, and the compromise multiplier for constructing comparison intervals—now gives us

$$(\text{H-spr}/1.349) \times \sqrt{(\pi/2)} \times (1.7/\sqrt{n}) = 1.58 \times \text{H-spr}/\sqrt{n}.$$

For further discussion of multiplicity and the statistical problem of multiple comparisons, the interested reader may consult the book by Miller (1966).

References

McGill, Robert, John W. Tukey, and Wayne A. Larsen. 1978. "Variations of Box Plots," *The American Statistician* 32:12–16.

Miller, Rupert G. 1966. *Simultaneous Statistical Inference*. New York: McGraw-Hill.

1976 Tax Guide for College Teachers. 1975. Washington, D.C.: Academic Information Service, Inc.

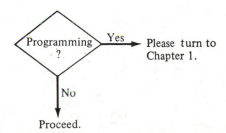

```
5000 REM   ONE OR THREE LINE BOXPLOT
5010 REM ENTRY CONDITIONS:M0,M9=MARGIN BOUNDS;
5020 REM V1 = VERSION:
5030 REM V1=1: 1-LINE BOXPLOT, V1=2 1-LINE NOTCHED BOXPLOT,
5040 REM V1=3: 3-LINE BOXPLOT, V1=4 3-LINE NOTCHED BOXPLOT
5050 REM V1<0 ASKS FOR DATA BOUNDS THEN USES ABS(V1) STYLE.
5060 REM C9 = # OF BOXES TO BE PRODUCED ON SAME SCALE
5070 REM IF C9 > 1, C() HOLDS GROUP ID'S.   THESE CAN
5080 REM BE ANY DISTINCT NUMBERS, BUT INTEGERS ARE BEST.
5090 REM BOXES WILL BE PRINTED IN GROUP ID ORDER.
5100 REM IF MULTIPLE BOXES PRINTED, Y() AND C() ARE SORTED ON C()
5110 REM IF DATA WERE NOT ORIGINALLY IN COLUMN-MAJOR ORDER,
5120 REM THIS CAN DESTROY CORRESPONDENCE WITH R() AND X().
5130 REM P9=# DESIRED POSITIONS;P()=CHR ARRAY;Y()=DATA ARRAY
5140 REM NICE #S SET AT 1,1.5,2,2.5,3,4,5,7,10
5150 REM OVERPRINTS WITH DECREASING PRECEDENCE:+=MEDIAN,
5160 REM ]=HI HINGE,[=LO HINGE,O=OUTSIDE OUTER FENCE,
5170 REM *=OUTSIDE INNER FENCE,|=EXTREMES,-=WHISKER
5180 REM POSITION FN =# CHRS TO RIGHT OF LEFT MARGIN
5185 REM
5190 REM   SORT Y() INTO W()

5200 GOSUB 3300
5210 IF V1 >= 0 GO TO 5290
5220 PRINT "MIN,MAX FOR BOXPLOT";
5230 INPUT L0,H1
5240 IF L0 < H1 THEN 5270
5250 PRINT L0;" IS NOT < ";H1;" RE-ENTER ";
5260 GO TO 5220
5270 LET V1 = ABS(V1)
5280 GO TO 5330
5290 REM

5300 REM   FIND NICE WIDTH

5310 LET H1 = W(N)
5320 LET L0 = W(1)
5330 LET N5 = 3
5340 LET P9 = M9 - M0 + 1
5350 LET A8 = 0

5360 REM   RETURNS P7=NPW

5370 GOSUB 1900

5380 REM   MULTIPLE BOXES?
5390 IF C9 <= 1 THEN 5750
```

82

```
5400 REM   YES, SORT INTO GROUP ID ORDER
5410 FOR I = 1 TO N
5420    LET W(I) = X(I)
5430    LET X(I) = C(I)
5440 NEXT I
5450 GOSUB 1200

5460 REM   X(), Y(), NOW SORTED BY GROUP ID

5470 FOR I = 1 TO N
5480    LET C(I) = X(I)
5490    LET X(I) = W(I)
5500 NEXT I

5510 REM   SAVE REAL N (COPYSORT WILL RESET IT)

5520 LET N7 = N

5530 REM   LEAVE ROOM TO LABEL BOXES. INTIGER ID #'S WORK BEST.

5540 LET M2 = LEN( STR$(C(N))) + 1
5550 IF M2 >= LEN( STR$(C(1))) THEN 5570
5560 LET M2 = LEN( STR$(C(1))) + 1
5570 LET M0 = M0 + M2
5580 LET J2 = 0

5590 REM   SET UP FOR THE NEXT ONE OF THE BOXES

5600 LET J1 = J2 + 1
5610 LET C7 = C(J1)
5620 LET C$ = STR$(C7)

5630 REM PRINT BOX LABEL ONLY IF THERE'S ROOM

5640 IF LEN(C$) > M2 THEN 5670
5650 PRINT TAB(M0 - M2);C$;

5660 REM FIND THE VALUES IN CURRENT BOX

5670 FOR J2 = J1 TO N7
5680    IF C(J2) <> C7 THEN 5710
5690 NEXT J2
5700 LET J2 = N7 + 1

5710 REM   COPY Y() FROM J1 TO J2 TO W() AND SORT
5715 LET J2 = J2 - 1
5720 GOSUB 3340

5730 REM   FIND MEDIAN(L1),HINGES(L2,L3),ADJACENT VALUE POINTERS(A1,A2)
5740 REM    FENCES(F1,F2), STEP(S1) OR DATA IN W()

5750 GOSUB 2500
5760 LET P2 = FNP(L2)
```

```
5770 LET P3 = FNP(L3)

5780 REM   WHICH STYLE BOX?

5790 IF V1 = 1 THEN 5930
5800 IF V1 = 3 THEN 5850

5810 REM   NOTCHED STYLE -- SET NOTCH BOUNDS AROUND MEDIAN

5820 LET X = 1.7 * (1.25 * (L3 - L2) / (1.35 * SQR(N)))
5830 LET N6 = FNP(L1 - X)
5840 LET N8 = FNP(L1 + X)
5850 IF V1 <= 2 THEN 5930

5860 REM   PRINT TOP OF BOX

5870 PRINT TAB(M0 + P2 - 1);
5880 IF P2 > P3 THEN 5920
5890 FOR I = P2 TO P3
5900    PRINT "-";
5910 NEXT I
5920 PRINT

5930 REM   CONSTRUCT LINE OF BOX IN PRINT ARRAY, P()
5940 REM   INITIALIZE P() TO BLANKS

5950 FOR I = 1 TO P9 + 1
5960    LET P(I) = ASC(" ")
5970 NEXT I

5980 REM   MARK LO WHISKERS, IF ANY

5990 IF FNP(W(A1)) > P2 - 1 THEN 6030
6000 FOR I = FNP(W(A1)) TO P2 - 1
6010    LET P(I) = ASC("-")
6020 NEXT I

6030 REM   MARK HI WHISKERS
6040 REM PROTECT US FROM UN-ANSI BASICS

6050 IF P3 + 1 > FNP(W(A2)) THEN 6090
6060 FOR I = P3 + 1 TO FNP(W(A2))
6070    LET P(I) = ASC("-")
6080 NEXT I

6090 REM   MARK EXTREMES

6100 LET P(1) = ASC("|")
6110 LET P9 = M9 - M0 + 1
6120 LET P(P9) = ASC("|")
```

```
6130 REM   MARK LO OUTLIERS, IF ANY

6140 IF A1 = 1 THEN 6220
6150 FOR I = 1 TO A1 - 1
6160    IF W(I) <= F1 - S1 THEN 6200
6170    IF W(I) > F1 THEN 6210
6180    LET P( FNP(W(I))) = ASC("*")
6190    GO TO 6210
6200    LET P( FNP(W(I))) = ASC("O")
6210 NEXT I

6220 REM   MARK HI OUTLIERS, IF ANY

6230 IF A2 = N THEN 6310
6240 FOR I = A2 + 1 TO N
6250    IF W(I) >= F2 + S1 THEN 6290
6260    IF W(I) < F2 THEN 6300
6270    LET P( FNP(W(I))) = ASC("*")
6280    GO TO 6300
6290    LET P( FNP(W(I))) = ASC("O")
6300 NEXT I

6310 REM   MARK HINGES

6320 LET P(P2) = 91
6330 LET P(P3) = ASC("]")
6340 IF V1 = 1 THEN 6390
6350 IF V1 = 3 THEN 6390

6360 REM   MARK NOTCHES

6370 LET P(N6) = ASC(">")
6380 LET P(N8) = ASC("<")

6390 REM   MARK MEDIAN

6400 LET P( FNP(L1)) = ASC("+")

6410 REM   NOW PRINT BOXPLOT
6420 REM   THERE MAY BE MORE EFFICIENT WAYS TO DO THIS ON SOME BASICS.

6430 PRINT TAB(M0);
6440 FOR I = 1 TO P9 + 1
6450    PRINT CHR$(P(I));
6460 NEXT I
6470 PRINT
6480 IF V1 <= 2 THEN 6560

6490 REM   PRINT THE BOTTOM OF THE BOX

6500 PRINT TAB(M0 + P2 - 1);
6510 IF P2 > P3 THEN 6560
```

```
6520 FOR I = P2 TO P3
6530    PRINT "-";
6540 NEXT I
6550 PRINT
6560 IF C9 <= 1 THEN 6620

6570 REM   MORE BOXES TO PRINT?

6580 IF J2 < N7 THEN 5600

6590 REM   NO, RESTORE N AND LEFT MARGIN

6600 LET N = N7
6610 LET M0 = M0 - M2
6620 RETURN
6630 END
```

```
      SUBROUTINE BOXES(Y, N, GSUB, NG, LINE3, NOTCH, SORTY, ERR)
C
C PRINT ADJACENT BOXPLOTS ON A SINGLE SCALE FOR ALL VARIABLES IN Y().
C
      INTEGER N, NG, ERR
      INTEGER GSUB(N)
      REAL Y(N), SORTY(N)
      LOGICAL LINE3, NOTCH
C
C Y()  CONTAINS DATA.  GSUB()  CONTAINS INTEGERS BETWEEN 1 AND NG
C IDENTIFYING THE DATA SET EACH ELEMENT OF Y() BELONGS TO.
C THIS DATA STRUCTURE IS CONSISTENT WITH THE SPARSE MATRIX FORMAT
C USED FOR STORING MATRICES IN OTHER PROGRAMS.  THE USE OF
C THE VECTOR  GSUB()  IS MEANT TO SUGGEST BOXPLOTS OF EITHER THE
C ROWS OR THE COLUMNS A MATRIX STORED IN THIS MANNER.
C IF  LINE3  IS .TRUE. ALL BOXPLOTS WILL BE FULL 3-LINE BOXPLOTS.
C IF  LINE3  IS .FALSE., ONE-LINE BOXPLOTS WILL BE PRINTED.
C  SCALING OF THESE PLOTS IS TO THE EXTREMES OF THE ENTIRE COMBINED
C DATA BATCH.  THE DETAILS OF EACH BOX, INCLUDING OUTLIER
C IDENTIFICATION, ARE DETERMINED FOR EACH BATCH INDIVIDUALLY.
C
C
      COMMON/CHRBUF/P, PMAX, PMIN, OUTPTR, MAXPTR, OUNIT
      INTEGER P(130), PMAX, PMIN, OUTPTR, MAXPTR, OUNIT
C
C LOCAL VARIABLES
C
      INTEGER NN, NPMAX, NPOS, LPMIN, SPMIN
      INTEGER CHRPAR, LBLW, OPOS, I, J, K
      REAL NICNOS(9), FRACT, UNIT, NPW, LO, HI
C
C
C FUNCTIONS
C
      INTEGER WDTHOF
C
C CALLS SUBROUTINES BOXP, NPOSW, PUTCHR, PUTNUM
C
      DATA NN,NICNOS(1),NICNOS(2),NICNOS(3)/9,1.0,1.5,2.0/
      DATA NICNOS(4),NICNOS(5),NICNOS(6)/2.5,3.0,4.0/
      DATA NICNOS(7),NICNOS(8),NICNOS(9)/5.0,7.0,10.0/
      DATA CHRPAR/44/
C
```

```
C   CHECK FOR AT LEAST 2 DATA VALUES.   OTHERWISE HIGHEST AND LOWEST
C   WILL BE EQUAL AND PLOT SCALING WILL FAIL ANYWAY.
C
        IF(N .GT. 1) GO TO 5
        ERR = 31
        GO TO 999
     5 LPMIN = PMIN + 7
        LO = Y(1)
        HI = Y(N)
        DO 10 I = 1, N
          IF(LO .GT. Y(I)) LO = Y(I)
          IF(HI .LT. Y(I)) HI = Y(I)
    10 CONTINUE
C
C   SCALE TO THE EXTREMES
C
        NPMAX = PMAX - LPMIN+1
        CALL NPOSW(HI, LO, NICNOS, NN, NPMAX, .FALSE., NPOS, FRACT,
       1    UNIT, NPW, ERR)
        IF (ERR .NE. 0) GO TO 999
C
C   NOW PRINT ALL THE BOXES.
C   DATA SETS ARE IDENTIFIED BY THEIR CODES IN GSUB()
C
        IF (NG .GT. 1) GO TO 17
        DO 15 K = 1, N
          SORTY(K) = Y(K)
    15 CONTINUE
        CALL BOXP(SORTY, N, LINE3, NOTCH, LO, HI, NPW, ERR)
        GO TO 999
    17 SPMIN = PMIN
        DO 30 I = 1, NG
          K = 0
          DO 20 J = 1, N
            IF(GSUB(J) .NE. I) GO TO 20
            K = K+1
            SORTY(K) = Y(J)
    20    CONTINUE
        PMIN = SPMIN
        LBLW = WDTHOF(I)
        OPOS = PMIN + 5 - LBLW
        CALL PUTNUM(OPOS, I, LBLW, ERR)
        OPOS = PMIN + 6
        CALL PUTCHR(OPOS, CHRPAR, ERR)
        IF(ERR .NE. 0) GO TO 999
        PMIN = LPMIN
        CALL BOXP(SORTY, K, LINE3, NOTCH, LO, HI, NPW, ERR)
        IF(ERR .NE. 0) GO TO 999
    30 CONTINUE
        PMIN = SPMIN
   999 RETURN
        END
```

```
      SUBROUTINE BOXP(SORTY, N, LINE3, NOTCH, LO, HI, NPW, ERR)
C
C
C   PRINT A BOXPLOT OF THE DATA IN SORTY()
C
C
      INTEGER N, ERR
      REAL SORTY(N), LO, HI, NPW
      LOGICAL LINE3, NOTCH
C
C   PLOT SCALING HAS BEEN DONE BY THE CALLING PROGRAM WITH NEEDED
C   INFORMATION PASSED IN AS   LO  (THE LOW EXTREME),   HI  (THE HIGH
C   EXTREME) AND  NPW  (THE NICE POSITION WIDTH FOR PLOTTING).
C   TYPICALLY THIS WILL BE ONE OF SEVERAL BOXPLOTS SCALED AND PRINTED
C   TOGETHER.
C      IF  LINE3  IS .TRUE. A 3-LINE BOXPLOT (FULL BOXES) IS PRINTED.
C   IF NOT, THE SIMPLE ONE-LINE BOXPLOT IS PRINTED.  BOTH CONVEY THE
C   SAME INFORMATION, BUT THE 3-LINE VERSION MAY LOOK NICER.
C      IF  NOTCH  IS .TRUE. A CONFIDENCE INTERVAL AROUND THE MEDIAN IS
C   INDICATED WITH PARENTHESES.
C
      COMMON/CHRBUF/P, PMAX, PMIN, OUTPTR, MAXPTR, OUNIT
      INTEGER P(130), PMAX, PMIN, OUTPTR, MAXPTR, OUNIT
C
C   FUNCTIONS
C
      INTEGER PLTPOS
C
C   CALL SUBROUTINES BOXTOP, PRINT, PUTCHR, YINFO
C
C   LOCAL VARIABLES
C
      INTEGER I, IADJL, IADJH, IFROM, ITO, LPMAX, LPMIN
      INTEGER OPOS, CHI, CHO, CHSTAR, CHMIN, CHPLUS, CHRPAR, CHLPAR
      REAL MED, HL, HH, ADJL, ADJH, STEP
      REAL FLOATN, NSTEP, LNOTCH, HNOTCH, OFENCL, OFENCH
C
      DATA CHI, CHO, CHPLUS, CHMIN, CHSTAR/9, 15, 39, 40, 41/
      DATA CHLPAR, CHRPAR/43, 44/
C
      LPMAX = PMAX
      LPMIN = PMIN
      CALL YINFO(SORTY, N, MED, HL, HH, ADJL, ADJH, IADJL, IADJH,
     1     STEP, ERR)
      IF (ERR .NE. 0) GO TO 999
      FLOATN = FLOAT(N)
      NSTEP = 1.7 * (1.25*(HH - HL)/(1.35* SQRT(FLOATN)))
      LNOTCH = MED - NSTEP
      HNOTCH = MED + NSTEP
C   PRINT TOP OF BOX, IF 3-LINE VERSION
      IF(LINE3) CALL BOXTOP(LO, HI, HL, HH, NPW, ERR)
      IF(ERR .NE. 0) GO TO 999
```

```
C
C   FILL CENTER LINE OF DISPLAY -- NOTE CAREFUL HIERARCHY
C   OF OVERPRINTING.  LAST PLACED CHARACTER IS ONLY ONE TO APPEAR.
C
C   MARK WHISKERS
C
      IFROM = PLTPOS(ADJL, LO, HI, NPW, ERR)
      ITO = PLTPOS(HL, LO, HI, NPW, ERR) - 1
      IF (IFROM .GT. ITO) GO TO 21
      DO 20 I = IFROM, ITO
        CALL PUTCHR(I, CHMIN, ERR)
   20 CONTINUE
   21 CONTINUE
      IFROM = PLTPOS(HH, LO, HI, NPW, ERR)
      ITO = PLTPOS(ADJH, LO, HI, NPW, ERR)
      IF (IFROM .GT. ITO) GO TO 31
      DO 30 I = IFROM, ITO
        CALL PUTCHR(I, CHMIN, EPR)
   30 CONTINUE
   31 CONTINUE
C
C   MARK LOW OUTLIERS, IF ANY
C
      IF(IADJL .EQ. 1) GO TO 41
      OFENCL = HL - 2.0*STEP
      ITO = IADJL - 1
      DO 40 I = 1, ITO
        OPOS = PLTPOS(SORTY(I), LO, HI, NPW, ERR)
        IF(SORTY(I) .LT. OFENCL) CALL PUTCHR(OPOS, CHO, ERR)
        IF(SORTY(I) .GE. OFENCL) CALL PUTCHR(OPOS, CHSTAR, ERR)
   40 CONTINUE
   41 CONTINUE
C
C   MARK HIGH OUTLIERS, IF ANY
C
      IF(IADJH .EQ. N) GO TO 51
      OFENCH = HH + 2.0*STEP
      IFROM = IADJH + 1
      DO 50 I = IFROM, N
        OPOS = PLTPOS(SORTY(I), LO, HI, NPW, ERR)
        IF(SORTY(I) .GT. OFENCH) CALL PUTCHR(OPOS, CHO, ERR)
        IF(SORTY(I) .LE. OFENCH) CALL PUTCHR(OPOS, CHSTAR, ERR)
   50 CONTINUE
   51 CONTINUE
```

```
C
C   MARK HINGES, NOTCHES, AND MEDIAN
C
      OPOS = PLTPOS(HL, LO, HI, NPW, ERR)
      CALL PUTCHR(OPOS, CHI, ERP)
      OPOS = PLTPOS(HH, LO, HI, NPW, ERR)
      CALL PUTCHR(OPOS, CHI, ERP)
      OPOS = PLTPOS(LNOTCH, LO, HI, NPW, ERR)
      IF(NOTCH) CALL PUTCHR(OPOS, CHLPAR, ERR)
      OPOS = PLTPOS(HNOTCH, LO, HI, NPW, ERR)
      IF(NOTCH) CALL PUTCHR(OPOS, CHRPAR, ERR)
      OPOS = PLTPOS(MED, LO, HI, NPW, ERR)
      CALL PUTCHR (OPOS, CHPLUS, ERR)
C
C   AND PRINT THE BOXPLOT
C
      IF(ERR .NE. 0) GO TO 999
      CALL PRINT
C
C   PRINT THE BOTTOM OF THE BOX
C
      IF(LINE3) CALL BOXTOP(LO, HI, HL, HH, NPW, ERR)
  999 RETURN
      END


      SUBROUTINE BOXTOP(LO, HI, HL, HH, NPW, ERR)
C
      REAL LO, HI, HL, HH, NPW
      INTEGER ERR
C
C   PRINT THE TOP OR BOTTOM OF A BOXPLOT DISPLAY
C
C   HI  AND  LO  ARE EDGES OF THE PLOTTING REGION USED BY THE PLTPOS
C   FUNCTION.
C   HL  AND  HH ARE THE LOW AND HIGH HINGES
C   NPW  IS THE NICE POSITION WIDTH SET BY THE PLOT SCALING ROUTINES
C
C   LOCAL VARIABLES
C
      INTEGER I, IFROM, ITO, CHMIN
C
C   FUNCTION
C
      INTEGER PLTPOS
C
C   DATA
C
      DATA CHMIN/40/
```

```
C
      IFROM = PLTPOS(HL, LO, HI, NPW, ERR)
      ITO = PLTPOS(HH, LO, HI, NPW, ERR)
      IF (IFROM .GT. ITO) GO TO 11
      DO 10 I = IFROM, ITO
        CALL PUTCHR(I, CHMIN, ERR)
   10 CONTINUE
   11 CONTINUE
      IF (ERR .EQ. 0) CALL PRINT
  999 RETURN
      END



      INTEGER FUNCTION PLTPOS(X, LO, HI, NPW, ERR)
C
C  FIND THE POSITION CORRESPONDING TO X ON PLOT BOUNDED
C  BETWEEN LO AND HI AND SCALED ACCORDING TO NPW.
C
      REAL X, LO, HI, NPW
      INTEGER ERR
C
C  FUNCTIONS
C
      INTEGER INTFN
C
C  COMMON
C
      COMMON /CHRBUF/P, PMAX, PMIN, OUTPTR, MAXPTR, OUNIT
      INTEGER P(130), PMAX, PMIN, OUTPTR, MAXPTR, OUNIT
C
      PLTPOS = INTFN((X-LO)/NPW, ERR) + PMIN
      IF (PLTPOS .LT. PMIN) PLTPOS = PMIN
      IF (PLTPOS .GT. PMAX) PLTPOS = PMAX
      RETURN
      END
```

Chapter 4

$x - y$ Plotting

Data that come as paired observations are usually displayed by drawing an x-y plot. This is a very common procedure and a powerful exploratory data-analysis tool. Plots of y versus x show at a glance how x and y are related to each other. For example, if larger y-values are often paired with larger x-values and smaller y-values with smaller x-values, that association will be evident in the plot. If the x-y points fall on or near a straight line, that will be clear from the plot—and we may be able to say more about the relationship between x and y, as we will see in Chapter 5. If the pattern of the plot shows a smooth change in y-values as we move from each x-value to the next larger one, we may want to look for a smooth pattern with techniques discussed in Chapter 6. And, as always, we will check the plot for any extraordinary points that do not seem to fit whatever pattern is present, for these points may deserve special attention.

ordered pair
    x-y data are often presented as **ordered pairs,** (x, y)—one ordered pair for each observation. Alternatively, such data can come as a pair of columns of numbers—one column for the x-values and one for the corresponding y-values. Such columns, whose values are in an established order (in this case, paired with each other), are examples of **arrays.** To refer to the ith value in an array,
array
subscript
we attach the **subscript** i to the name of the array; for example, x_i. The ith x–y observation is (x_i, y_i).

93

Exhibit 4-1   Births per 10,000 23-Year-Old Women in the United States from 1917 to 1975

Year	Birthrate	Year	Birthrate
1917	183.1	1946	189.7
1918	183.9	1947	212.0
1919	163.1	1948	200.4
1920	179.5	1949	201.8
1921	181.4	1950	200.7
1922	173.4	1951	215.6
1923	167.6	1952	222.5
1924	177.4	1953	231.5
1925	171.7	1954	237.9
1926	170.1	1955	244.0
1927	163.7	1956	259.4
1928	151.9	1957	268.8
1929	145.4	1958	264.3
1930	145.0	1959	264.5
1931	138.9	1960	268.1
1932	131.5	1961	264.0
1933	125.7	1962	252.8
1934	129.5	1963	240.0
1935	129.6	1964	229.1
1936	129.5	1965	204.8
1937	132.2	1966	193.3
1938	134.1	1967	179.0
1939	132.1	1968	178.1
1940	137.4	1969	181.1
1941	148.1	1970	165.6
1942	174.1	1971	159.8
1943	174.7	1972	136.1
1944	156.7	1973	126.3
1945	143.3	1974	123.3
		1975	118.5

Source: P.K. Whelpton and A.A. Campbell, "Fertility Tables for Birth Charts of American Women," *Vital Statistics—Special Reports* 51, no. 1 (Washington, D.C.: Government Printing Office, 1960) years 1917–1957. National Center for Health Statistics, *Vital Statistics of the United States Vol. 1, Natality* (Washington, D.C.: Government Printing Office, yearly, 1958–1975).

4.1   *x-y* Plots

x-y plots are common in books and magazines, so we consider them only briefly. We recall that each point on the plot is located simultaneously by its position on the horizontal *x*-axis (corresponding to its value on the *x*-variable) and by its position on the vertical *y*-axis (corresponding to its value on the *y*-variable).

For example, Exhibit 4–1 lists the number of live births per 10,000 23-year-old women in the United States between 1917 and 1975. To examine patterns in the birthrate over time, we plot birthrate (*y*) on the vertical axis against year (*x*) on the horizontal axis. The hand-drawn result is shown in Exhibit 4–2. Each point on the plot can be easily matched with its pair of data values by finding the numbers associated with its position on each axis. The global pattern in the plot shows that the birthrate fell sharply during the 1920s, bottomed out during the Depression, rose rapidly to a peak around 1960, and has fallen rapidly since then.

Although there is little to say about hand-drawn exploratory *x-y* plots, there is much to consider when the computer prints the plot. The remainder of this chapter is devoted to computer-produced *x-y* plots—and primarily to a

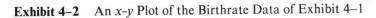

Exhibit 4–2   An *x-y* Plot of the Birthrate Data of Exhibit 4–1

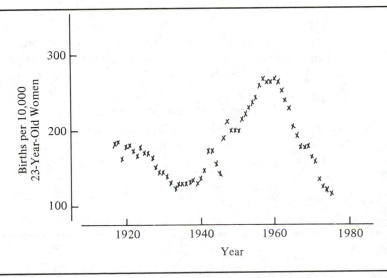

particular type of plot designed for exploratory data analysis and for interactive computing on a standard typewriter-style computer terminal. If you do not intend to use a computer in your exploratory analyses, you can skip the rest of this chapter without any loss of continuity. If your computer system is already equipped with some other version of x-y plotting (as it will almost certainly be if you are using a statistical package), you may prefer to substitute that version for the method presented here. Nevertheless, you should read the rest of this chapter because it includes fundamental ideas about computer-printed plots and provides a useful background for anyone using the computer to print x-y plots.

4.2   Computer Plots

Most computer programs for x-y plots concentrate on making them nice in some chosen way. The programs presented here concentrate on making the plot concise, so that it can be generated quickly on a computer terminal, and on making the scaling and labeling of the plot natural and close to what we might choose if we were drawing it by hand.

In drawing a plot by hand, we can place points exactly where they belong, guided by the ruled grid lines of the graph paper. A point can fall on a grid line or anywhere between the sets of lines. However, computer terminals are usually limited to choosing a character position across the line to represent the x-coordinate, choosing a print line on the page to represent the y-coordinate, and printing a character at that location. We may think of such a computer plot as being drawn on graph paper on which each box of the grid must either be entirely colored in or left blank. To make matters worse, the boxes are not even square, since printing characters are usually about twice as tall as they are wide. Nevertheless, such plots can be made easy to read and are valuable ways to display data. Exhibit 4–3 shows a fairly typical computer-terminal plot of the birthrate in Exhibit 4–1 with the character 0 as the plotting symbol.

4.3   Condensed Plots

Since computer plots must use either all of a "character box" or none of it, we are tempted to make the plots large so that each character box will have a

Exhibit 4–3   A Computer-Produced Plot of the Birthrate Data of Exhibit 4–1

```
+ 268                                          0  0
+ 264                                           00 0
+ 260
+ 256                                        0
+ 252                                            0
+ 248
+ 244                                      0
+ 240                                          0
+ 236                                      0
+ 232
+ 228                                    0       0
+ 224
+ 220                                    0
+ 216
+ 212                                0   0
+ 208
+ 204                                          0
+ 200                               000
+ 196
+ 192                                          0
+ 188                             0
+ 184
+ 180   00  0                                0
+ 176     0   0                               00
+ 172      0                0 0
+ 168        00
+ 164      0                                 0
+ 160    0      0
+ 156                        0               0
+ 152
+ 148         0        0
+ 144          00
+ 140                    0
+ 136           0      0               0
+ 132              000
+ 128           0 000
+ 124             0                     0
+ 120                                   0
+ 116                                    0
        ---+----------+----------+----------+----------+----------+----------+----------+----------+
```

more precise meaning and thus give the plot greater resolution. Unfortunately, large plots are very slow to print on most interactive computer terminals. This slowness can be a major handicap in exploratory data analysis because we might want to look at several plots or at slightly different versions of the same plot. Therefore, we seek a way to condense an *x-y* plot so that it will take less space and print faster without sacrificing precision. The simple choice available is the selection of the character used to mark a box as filled.

We can condense the plot vertically by squeezing as many as 10 lines of plot into a single line and using the printed character—say, a numeral from 0 to 9—to indicate the original line occupied by the point. This device reproduces the plot in $\frac{1}{10}$ the original number of lines (typically down from 50 or 60 lines to 5 or 6 lines) with surprisingly little loss of precision. The improvement is so great that we can afford to be a bit greedy and use 10 lines or so and obtain a plot that contains, though unobtrusively, even more information than we displayed originally.

4.4   Coded Plot Symbols

In implementing condensed plots, we choose to number the subdivisions of each line according to their distance from zero, with 0 labeling the subdivision nearest zero and 9 the subdivision farthest from zero. Thus, for positive *y*-values on the same print line, 9 indicates a point higher than a point labeled 8, while for negative *y*-values a point labeled 9 will be lower than a point on the same line labeled 8. Exhibit 4–4 illustrates the condensation in plotting the birthrate data.

Comparing the two plots in Exhibits 4–3 and 4–4 shows how condensing the plot uses digits to convey information about the data points. As an example of the details, let us see what happens to the first point, (1917, 183.1), and the fifth point, (1921, 181.4), in these plots. In Exhibit 4–2 we could indicate the values of these two points fairly closely. However, the computer-produced plot in Exhibit 4–3 tells us only that their *y*-values fall in the interval $180 \leq y < 184$. In Exhibit 4–4, even though it uses only about one-fifth as many lines, these two points are represented by the symbols 1 and 0, respectively, on the line labeled $+180$. Because we are using 10 characters (0 through 9) per line, we know that the *y*-value of the first point falls in the second tenth of the interval $180 \leq y < 200$—that is, between 182 and 184. Similarly, the *y*-value of the fifth point is in the first tenth, between 180 and 182.

Exhibit 4–4   A Condensed Plot of the Birthrate Data of Exhibit 4–1

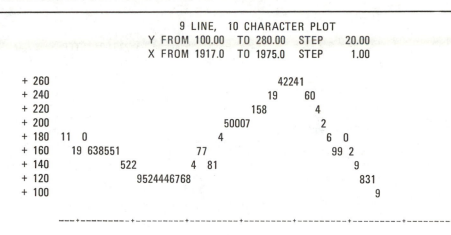

```
                              9 LINE,  10 CHARACTER PLOT
                        Y FROM 100.00  TO 280.00  STEP    20.00
                        X FROM 1917.0  TO 1975.0  STEP     1.00

        + 260                                   42241
        + 240                               19      60
        + 220                           158       4
        + 200                       50007       2
        + 180  11  0                  4            6  0
        + 160     19 638551         77           99 2
        + 140           522       4  81                9
        + 120        9524446768                     831
        + 100                                         9


        ---+----------+----------+----------+----------+----------+----------+
```

Of course, in condensing the *y*-axis, we sacrifice some things to gain speed and conciseness. First, patterns immediately visible in a full-page plot may be a little harder to see in the 10-line version, although experience has shown that most patterns are still clear even without reading the digits for fine details. Second, we simultaneously make overprints—that is, two or more points falling in the same box—more likely and harder to indicate. (Some plotting programs indicate overprints with different characters, often numerals!) This second sacrifice is usually acceptable for exploratory analyses. Third, the use of 10 characters may add too much confusion to an already complex plot. We can remedy this confusion by allowing the choice of fewer subdivisions of each line; the programs allow any choice between 1 and 10 numeric codes.

Since the problems of condensed plotting increase as we condense to fewer lines while the benefits of speed and smaller size increase, the choice of numbers of lines and characters is best left to the user's discretion, so that the correct balance can be struck for any particular data set or any particular computer terminal. Condensed plots begin with a legend:

9 LINE, 10 CHARACTER PLOT

Y FROM 100.0	TO 280.0	STEP	20.0
X FROM 1917	TO 1975	STEP	1.0

Exhibit 4–5   A 6-Line, 4-Character Plot of the Birthrate Data of Exhibit 4–1

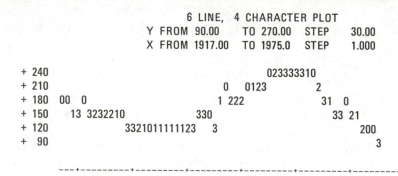

```
                              6 LINE,  4 CHARACTER PLOT
                       Y FROM 90.00     TO 270.00  STEP    30.00
                       X FROM 1917.00   TO 1975.0  STEP    1.000


+ 240                                          023333310
+ 210                               0    0123          2
+ 180   00  0                        1 222              31  0
+ 150      13 3232210                 330                    33 21
+ 120            3321011111123    3                           200
+  90                                                          3


     ---+---------+---------+---------+---------+---------+---------+---------+
```

The legend tells how many lines the plot actually requires and how finely the lines are subdivided—that is, the number of characters. It then reports the extent of the data values accommodated by the entire plot and the range of data values accommodated by each line (*y* STEP) and by each horizontal character position (*x* STEP). Together, these make it easy to determine the magnitude of the data values (the *y*-axis labels do not include decimal points) and to translate any particular plotted point into its numeric value. Because the *y*-axis labels report the value of the inner (near zero) edge of each line, the *y*-bounds reported in the legend will typically extend beyond the outer axis labels. Note that a 40-line, 1-character plot is essentially the standard *x-y* plot made on a computer terminal. Indeed, that is how Exhibit 4–3 was generated. Exhibit 4–5 shows a 6-line, 4-character plot of the birthrate data. This form of the display was originally proposed by Andrews and Tukey (1973).

4.5   Condensed Plots and Stem-and-Leaf Displays

Astute readers may have noticed a resemblance between condensed plots and stem-and-leaf displays. The *y*-axis labels are similar to stems, and the characters chosen to provide additional information about the *y*-values are much like leaves. All we have done is stretch the leaves across the page according to the value of some other variable represented on the *x*-axis.

Indeed, the algorithms to generate these displays are quite similar. Of course, the numerals used in plotting are often not exactly like leaves because they may not represent a specific digit of the *y*-value but rather a subdivision of the line.

For example, Exhibit 4–6 shows the precipitation pH data that we have analyzed in previous chapters and the date of the precipitation recorded as day number in 1974, where dates in 1973 are negative and multiple-day

Exhibit 4–6   Precipitation pH and Day Number of Event (Jan. 1 = day 1. Multiple-day precipitation events are plotted at the average day number.)

Day No.	pH
−11	4.57
−5.5	5.62
−1	4.12
9	5.29
18.5	4.64
21	4.31
26.5	4.30
28	4.39
37.5	4.45
41	5.67
47.5	4.39
54.5	4.52
55.5	4.26
60	4.26
68	4.40
69	5.78
75.5	4.73
81	4.56
90	5.08
94.5	4.41
98.5	4.12
105	5.51
116.5	4.82
132.5	4.63
138	4.29
144	4.60

Note: Data from Exhibit 1–1.

Exhibit 4–7   Condensed Plot of Precipitation pH versus Day of 1974

```
                          10 LINE,  10 CHARACTER PLOT
                          4.05 < Y < 5.55,   STEP = .15
                         -12.5 < X < 145,    STEP = 2.5

        + 540   P                    P          P          7
        + 525        2
        + 510
        + 495                                      8
        + 480                                                  1
        + 465                                   5
        + 450   4         9           1       4            8   6
        + 435        2   6   2       3        4
        + 420      7 6           4 4                     6
        + 405   4                          4
```

precipitation events are plotted at the middle day of the event. Exhibit 4–7 shows the condensed plot. Compare this plot with the stem-and-leaf display of these data in Exhibit 1–10. The three outlying values identified by the stem-and-leaf program are represented by P's. There doesn't appear to be any strong pattern in this plot, although some increase in pH may have occurred after day 60 (1 Mar.).

The close similarity of stem-and-leaf displays and condensed plots provides insight into the plotting of negative y-values. Condensed plots use larger numbers to indicate points farther from zero on the same print line. As a result, increasing the numeric code moves points up on a positive line but down (away from zero) on a negative line. This is consistent with practice in a stem-and-leaf display, where larger leaves on negative stems indicate more negative (farther from zero) values.

Condensed plots may also have a line labeled -00 for the same reason that stem-and-leaf displays can have a -0 stem. Small negative values just below zero will naturally be plotted on the -00 line. (Review Section 1.3 for a discussion of this.)

Because the plotting symbols increase away from the level $y = 0$, it is important to know where this level is on the plot. When necessary (the algorithm in Section 4.8 specifies exactly when), this level is marked on the plot. The exact point where y equals 0 really falls between the two 00 lines, so it is indicated with symmetrically placed marks on both of these lines. The BASIC program begins the $+00$ and -00 lines with a "herringbone" that graphically points to the invisible x-axis running between these lines. It looks like this:

```
+00)\\\\\
-00)/////
```

FORTRAN lacks the backslash character (\), so its marker consists of parallel minus signs:

```
+00)-----
-00)-----
```

Any data value that should be plotted in one of the marked positions replaces the axis mark. Exhibit 4–8 shows an example, plotting the January temperature against the air pollution potential of hydrocarbons in 60 SMSAs. (See Exhibits 1–7 and 1–5 for the stem-and-leaf displays of the temperature and HC data.)

Finally we note that, as in the stem-and-leaf display, y-values exactly equal to zero do not clearly belong on either the $+00$ or the -00 line. (Or, more properly, they belong on both.) In the stem-and-leaf display, we split zeros between the two middle lines, but splitting in this way could disturb patterns in an x-y plot. Here the usual rule is to assign zeros to the $+00$ line. However, if the data contain no positive values, we place the zero values on the -00 line. Handling this special case in this way saves a plot line and avoids separating zero values from small negative values.

Exhibit 4–8   January Temperature (°C) versus Air Pollution Potential of Hydrocarbons in 60 SMSAs

```
                        9 LINE,  10 CHARACTER PLOT
                  Y FROM   -12.0  TO   15.0   STEP   3.0
                  X FROM     1.0  TO   66.0   STEP   1.0


        + 120    P  2              0                                      R
        +  90                                                            R
        +  60   4               4             4                          R
        +  30         4   2  8 48 4                         1
        +  00)  ---0   1  0  1          3  0 0     5  1 5        9
        -  00)  ---3759    957  3   9 1      7          5      9
        -  30         6 46 4      4                                      R
        -  60                           2
        -  90                    7

        ---------+---------+---------+---------+---------+---------+---------+----
```

4.6   Bounds for Plots

data bounds

Whenever we display data graphically, we must decide whether to plot every number or exclude possible outliers so they do not dominate the display. The condensed plotting programs automatically exclude values beyond the fences, just as the stem-and-leaf programs do. Now, of course, we need to know the *data bounds* in both the x and y directions. (See Appendix A for the technical details of these decisions.)

Because the plot is adjusted to be easy to read and to include all the points within the data bounds, it is likely that the actual edges of the plot will be slightly beyond the data bounds. These bounds are printed above the plot in the legend.

Numbers that fall outside the plot bounds are indicated with special characters along the edges of the plot, as described by the following diagram:

*		P		*
L		PLOT		R
*		M		*

That is, points whose y-values are too high appear as a P (for "plus") on the top line of the plot at the horizontal position appropriate for their x-value. Similarly, points with extremely low y-values appear on the bottom line of the plot as an M (for "minus"). Points outside the horizontal plot bounds appear, on the line corresponding to their y-value, in the leftmost or rightmost position as an L (left) or R (right). Points that are extreme in two directions appear in a corner position of the plot as an asterisk (*).

Exhibit 4–7 shows such data bounding in the y-axis dimension, and Exhibit 4–8 shows bounding in both dimensions. In the second case especially, the exclusion of cities with extraordinarily large hydrocarbon air pollution potentials has preserved the patterns in the display. To see this, recall from Exhibit 1–8 how extreme the high hydrocarbon values are. If we had tried to include Los Angeles (at 648) on the plot, most of the other points would have been hopelessly crowded to the left.

Whenever fewer than 10 characters are being used for plotting, the unused "improper" characters are used on the highest and lowest lines to indicate points just beyond the plot bounds. For example, on a 6-line, 8-character (0 through 7) plot, a point just barely too high for the top line will appear on that line as the "improper" digit 8. Had this been an 8-line plot with the same scaling, this point would have appeared on the next higher line as a 0. Similarly, a point just far enough above this last one to require a new digit—that is, a point that would have appeared as a 1 on the next higher line had there been one—will appear as a 9. Points too far away from the plot center to be represented with improper digits are plotted with M and P. This coding provides precise information about the location of points printed for such "near outliers" will indicate how many lines they are beyond the edge of the plot. Thus, a 2 says that the point is on the second line beyond the lines now printed.

4.7   Focusing Plots

Although the condensed plotting programs provide default choices of data bounds, at times it is useful to override these choices. The plotting programs can be focused on any region of the *x-y* plane by specifying minimum and maximum values for each axis. If the data extremes are specified, the plot will include all of the data points. If a small region is selected, this region will be blown up to fill the entire plotting area, and points beyond the specified borders of that region will be treated as outliers. This feature makes it possible to focus on a portion of a complex display so as to better understand its fine structure.

It is also possible to divide part of the *x-y* plane into equal-sized rectangular regions and to generate condensed plots for each region (or just for regions known to contain data points). These plots can then be pasted together to obtain a highly precise montage display. If the regions are the same size, the plots will have the same scale. With practice, the top and bottom plot lines, which will fill with "outliers," can be made superfluous by overlapping the regions slightly. For example, five 10-line, 10-character plots can be used to cover a smoothly increasing relationship by choosing regions placed diagonally across the *x-y* plane. The resulting montage will have the same vertical resolution as a 500-line printer plot—close to the resolution possible on many graphics devices—yet the display will have taken only 50 lines and about 2 minutes (at 30 characters per second) to print.

4.8   Using the Programs

The condensed plot programs accept pairs of data values specified as corresponding elements of two arrays. For example, the first element of one array and the first element of the other array make up the first (x, y) pair. The number of lines and number of characters may be specified. If these are not specified, the program uses 10 lines and 10 characters. In addition, a choice is available between either plotting all the data or focusing only on data between the adjacent values on both x and y; the latter choice is the default. Alternatively, explicit bounds for x or y can be specified.

† 4.9   Algorithms

The design principles of the plotting algorithm are described in Appendix A, which should be read at this time. This section uses the vocabulary established in that appendix.

The programs accept data value pairs in arrays X( ) and Y( ). They find the adjacent values for both Y and X and use them to establish scale factors for each dimension. Because the scale factors are "nice" numbers, the viewport may extend beyond the adjacent values. The legend is printed first to identify the region of the x-y plane being displayed. Data in X and Y are ordered on Y, retaining the pairing. The programs then step through the y-values in much the same way as in the stem-and-leaf programs.

The plot is printed one line at a time. First, the y-label is constructed much as a stem, but with as many as four digits. Then, for the values on the current line, a plot symbol and x-position are determined. If the determined print position is already filled, the more extreme of the two plot symbols is retained. When all the data values belonging on that line have been processed, the line is printed. The programs note the print position of the rightmost point on the line, so that the line can be printed efficiently.

The $+00$ and -00 lines are marked to indicate the location of $y = 0$ if both positive and negative y-values are to be plotted and if the marked lines are at least three lines from the nearer edge of the display. The zero indicators

```
\\\\\
/////   in BASIC          -----
                          -----   in FORTRAN
```

are placed on the line first and replaced by any data points falling into those plot positions.

† 4.10   Alternatives

It is possible to produce plots that offer a compromise between precision and graphic impact by choosing plotting symbols that themselves contribute to the graphics. What is needed is a set of symbols that prints progressively higher on the line. One possible set is {_ -ˆ}. This scheme can easily go awry when the programs can be used from many different output terminals. (The example set given would become {← -↑} on some devices—far from the intended impression.)

More palatable alternatives are available to users with high-quality graphic devices. The resolution of many of these devices is 500 to 1000 vertical plot positions, which is far better than we can achieve with a condensed plot of reasonable size. Readers wishing to use such devices may want to use the plot-scaling programs provided in this book (see Appendices A and B). These programs can be modified easily to suit any plotting device, and they incorporate several features valuable in exploratory analyses. Appendix A discusses these features and their function in exploratory analysis.

† 4.11   Details of the Programs

FORTRAN

The FORTRAN subroutine PLOT is invoked with the statement

            CALL PLOT(Y, X, N, WY, WX, LINSET, CHRSET, XMIN, XMAX, YMIN, YMAX, ERR)

where

X( ) and Y( )	hold the N ordered pairs (X(i), Y(i));
N	is the number of data values;
WX( ) and WY( )	are N-long work arrays to hold the (x, y) values sorted on y;
LINSET	specifies the maximum number of lines to be used in the body of the plot. (The scaling routines may decide to use fewer lines.);

CHRSET	specifies the number of subdivisions (characters) of each line. It can be no greater than 10. If either LINSET or CHRSET is zero, the plot format defaults to 10 lines, 10 characters;
XMIN and XMAX	specify the range of *x*-values to be covered by the plot;
YMIN and YMAX	specify the range of *y*-values to be covered by the plot;
	For either pair of bounds, if the minimum and maximum bounds are equal, the program defaults to using adjacent values on that dimension;
ERR	is the error flag, whose values are

	0	normal
	41	N < 5—too few points to plot
	42	violates $5 \leq$ lines ≤ 40 or $1 \leq$ characters ≤ 10
	44	all *x*-values equal; no plot possible
	45	all *y*-values equal; no plot produced.

BASIC

The BASIC subroutine is entered with N data pairs (X(i), Y(i)) in arrays X( ) and Y( ). The plot format is specified by the version number, V1:

V1 = 1	6-line, 4-character (Andrews-Tukey) plot;
V1 = 2	10-line, 10 character plot;
V1 = 3	30-line, 1-character plot (ordinary computer plot);
V1 < 0	asks for input to override all scaling options.

All of the pre-set plots are scaled automatically to the adjacent values in both dimensions. The program builds each line of the plot in the P( ) vector so that overprints can be dealt with gracefully. Because the program stores the ASCII values of characters and numerals, the check performed to select the more extreme of two values falling at the same plot position depends on the ASCII collating sequence. Programmers on non-ASCII systems should check the indicated portions of the code to be sure the collating sequence that their systems use is compatible.

On small computers, sorting on Y( ) and carrying X( ) can be time-consuming. Time spent optimizing this subroutine for a particular machine can significantly improve the speed of the plotting programs.

Reference

Andrews, David F., and John W. Tukey. 1973. "Teletypewriter Plots for Data Analysis Can Be Fast: 6-line Plots, Including Probability Plots." *Applied Statistics* 22:192–202.

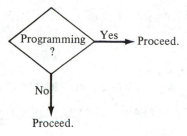

```
5000  REM    CONDENSED PLOTTING SUBROUTINE
5010  REM    PLOT Y() VS X(), LENGTH N
5020  REM    ON EXIT DATA IS RESORTED ON X() CARRYING Y().
5030  REM    VERSIONS: V1=1 : 6-LINE, 4-CHARACTER (ANDREWS-TUKEY) PLOT
5040  REM    V1=2 : 10-LINE, 10-CHARACTER PLOT
5050  REM    V1=3 : 30-LINE, 1-CHARACTER PLOT (OLD-STYLE PLOT)
5060  REM     V1<0 ASKS FOR INPUT TO OVERRIDE ALL SCALING OPTIONS.

5070  LET L = 6
5080  LET C = 4
5090  IF V1 = 1 THEN 5330
5100  LET L = 10
5110  LET C = 10
5120  IF V1 = 2 THEN 5330
5130  LET L = 30
5140  LET C = 1
5150  IF V1 = 3 THEN 5330
5160  IF V1 < 0 THEN 5190
5170  PRINT "ILLEGAL PLOT VERSION SPECIFIED:"

5180  REM   L=#LINES,C=#CHRS,Q$=DATA BOUND MODE OF OLD,NEW,DEFAULT

5190  PRINT TAB(M0);"#LINES,#CHRS";
5200  INPUT L,C
5210  PRINT "DATA BOUND MODE";
5220  INPUT Q$
5230  IF Q$ = "DEFAULT" THEN 5330

5240  REM    STILL NEED TO SORT EVEN IF NOT AUTO SCALING.
5250  REM    SORT ON Y CARRYING X

5260  GOSUB 1400
5270  GOSUB 1200
5280  GOSUB 1400
5290  IF Q$ = "NEW" THEN 5530
5300  IF Q$ = "OLD" THEN 5550
5310  PRINT TAB(M0);"DATA BOUND MODE MUST BE OLD, NEW, OR DEFAULT"
5320  GO TO 5210

5330  REM   GET DEFAULT LIMITS FOR X-Y PLOT IN P1,P2,P3,P4
5340  REM   COPY X() TO W() AND SORT

5350  GOSUB 3000
5360  GOSUB 2500
5370  LET P3 = A3
5380  LET P4 = A4
5390  IF P4 > P3 THEN 5420
5400  PRINT TAB(M0);"X-RANGE ZERO"
5410  STOP
```

110

```
5420 REM   SORT ON Y() CARRYING X() (UTILITY SORT DOES THE REVERSE).

5430 GOSUB 1400
5440 GOSUB 1200
5450 GOSUB 1400
5460 FOR I = 1 TO N
5470    LET W(I) = Y(I)
5480 NEXT I
5490 GOSUB 2500
5500 LET P1 = A4
5510 LET P2 = A3
5520 GO TO 5600
5530 PRINT TAB(M0);"DATA BOUNDS: TOP, BOTTOM, LEFT, RIGHT";
5540 INPUT P1,P2,P3,P4
5550 IF P1 > P2 THEN 5580
5560 PRINT TAB(M0);"ILLEGAL BOUNDS"
5570 GO TO 5190
5580 IF P3 >= P4 THEN 5560

5590 REM   SET UP MARGINS

5600 LET M = M9 - M0 - 5
5610 IF M >= 22 THEN 5640
5620 PRINT TAB(M0);"MARGIN BOUNDS ";M0;M9;" TOO SMALL A SPACE"
5630 STOP
5640 IF L > 0 THEN 5670
5650 PRINT TAB(M0);"1 TO 40 LINES, 1 TO 10 CHARACTERS"
5660 GO TO 5190
5670 IF L > 40 THEN 5650
5680 IF C > 10 THEN 5650
5690 IF C < 1 THEN 5650
5700 LET C = INT(C)

5710 REM   FIND A NICE LINE HEIGHT

5720 LET H1 = P1
5730 LET L0 = P2
5740 LET P9 = INT(L)
5750 LET N5 = 3
5760 LET A8 = 1
5770 GOSUB 1900

5780 REM   PRESERVE THE Y-DIRECTION UNIT

5790 LET U1 = U
5800 IF N4 <> 10 THEN 5850
5810 LET N4 = 1
5820 LET N3 = N3 + 1
5830 LET U1 = 10 ^ N3
```

```
5840 REM  L1=NICE LINE WIDTH,L=#LINES REQUIRED,L2=L/2 FOR FORMAT

5850 LET L1 = P7
5860 LET L = P8
5870 LET L2 = INT(L / 2)
5880 LET H1 = P4
5890 LET L0 = P3
5900 LET P9 = M
5910 LET A8 = 0
5920 GOSUB 1900
5930 LET M1 = P7
5940 LET M = P8

5950 REM   M1=NICE WIDTH OF 1 CHARACTER IN X,M=NICE MARGIN REQUIRED
5960 REM   DETERMINE NICE DATA BOUNDS
5970 REM   FIND NICE PLOT EDGES--ROUND AWAY FROM CENTER OF PLOT

5980 LET P2 = FNF(P2 / L1) * L1
5990 LET Y4 = FNC(P1 / L1)

6000 REM Y4 IS # LINES FROM ZERO. IT IS USED TO CONSTRUCT LINE LABELS
                                                            SAFELY

6010 LET P1 = Y4 * L1
6020 LET P3 = FNF(P3 / M1) * M1
6030 LET P4 = FNC(P4 / M1) * M1

6040 REM   NOW DATA BOUNDS ARE NICE

6050 PRINT TAB(M / 2 - 11);L;" LINE, ";C;" CHARACTER PLOT"
6060 PRINT
6070 PRINT TAB(M0);P2;"< Y <";P1;", STEP =";L1
6080 PRINT TAB(M0);P3;"< X <";P4;", STEP =";M1
6090 PRINT

6100 REM   INITIALIZE FOR PLOTTING:L5=LINE WIDTH MANTISSA FOR LABELS
6110 REM   Y2=CUT IN Y DIRECTION--STARTED ONE L1 TOO HIGH
6120 REM   Y3=EDGE OF LINE NEAREST 0,USED TO FIND CHARACTER
6130 REM   L8=LABEL;N7=POSITIVE FLAG;L9=LINE COUNT

6140 LET L5 = L1 / U1
6150 LET Y2 = P1
6160 LET Y3 = Y2
6170 IF Y2 >= 0 THEN 6190
6180 LET Y3 = Y2 + L1
6190 LET N7 = 1
6200 IF P1 >= 0 THEN 6220
6210 LET N7 = 0
6220 LET L9 = 0
6230 LET K = N + 1
```

```
6240 REM   START A NEW LINE OF PLOT

6250 FOR I = 1 TO M
6260    LET P(I) = ASC(" ")
6270 NEXT I
6280 LET P6 = 0

6290 REM POINTER TO PRINTING CHARACTER

6300 IF Y2 = 0 THEN 6320
6310 LET Y3 = Y3 - L1
6320 LET Y2 = Y2 - L1
6330 LET L9 = L9 + 1

6340 REM   PRINT THE LABEL TO START THE LINE

6350 LET Y4 = FNI(Y4 - 1)
6360 LET L8 = Y4 * L5
6370 ON SGN(Y4) + 2 GO TO 6390,6410,6670

6380 REM                        -     0     +

6390 PRINT TAB(M0);"-";
6400 GO TO 6700
6410 IF N7 = 0 THEN 6580
6420 PRINT TAB(M0);"+  00:";
6430 LET N7 = 0
6440 LET Y4 = FNI(Y4 + 1)

6450 REM   MARK ZERO LINES SINCE CHARACTERS COUNT OTHER WAY PAST HERE

6460 LET F3 = 0
6470 IF C = 1 THEN 6720
6480 IF L - L9 <= 2 THEN 6720
6490 LET F3 = 1

6500 REM ASCII BACK SLASH IS 92

6510 FOR I = 1 TO 5
6520    LET P(I) = 92
6530    LET P(M - I + 1) = ASC("/")
6540 NEXT I
6550 LET P6 = M
6560 GO TO 6720

6570 REM   -00 LINE

6580 PRINT TAB(M0);"-  00:";
6590 IF F3 <> 1 THEN 6720
6600 FOR I = 1 TO 5
6610    LET P(I) = ASC("/")
6620    LET P(M - I + 1) = 92
6630 NEXT I
```

```
6640 LET P6 = M
6650 GO TO 6720

6660 REM   POSITIVE LINE

6670 PRINT TAB(M0);"+";

6680 REM   THE 3 MOST INTERESTING DIGITS ARE EITHER SIDE OF THE ONE
6690 REM   POINTED TO BY THE UNIT.  USE THEM FOR Y LABEL.

6700 LET L$ = STR$( FNI(10 * ABS(L8)))
6710 PRINT TAB(M0 + 5 - LEN(L$));L$;":";

6720 REM   GET NEXT DATA POINT

6730 LET K = K - 1
6740 IF K <= 0 THEN 7200
6750 LET X7 = X(K)
6760 LET Y7 = Y(K)
6770 IF (1 + E0) * Y7 > = Y2 THEN 6830

6780 REM   LAST LINE SKIPS CHECK FOR NEXT LINE

6790 IF L9 = L THEN 6830
6800 LET K = K + 1

6810 REM   NEED A NEW LINE--WRAP THIS ONE UP

6820 GO TO 7210

6830 REM   GET CHARACTER FOR DETAIL ON Y POSITION

6840 LET Y0 = INT( ABS(((1 + E0) * Y7 - Y3) / L1) * C)

6850 REM   Y0 IS THE NUMBER TO PRINT

6860 LET Y1 = ASC("0") + Y0
6870 IF Y0 <= 9 THEN 6910
6880 LET Y1 = ASC("M")
6890 IF L9 = L THEN 6910
6900 LET Y1 = ASC("P")

6910 REM   GET X POSITION AND PLACE CHARACTER THERE

6920 LET X0 = FNI((X7 - P3) / M1) + 1
6930 IF X0 >= 1 THEN 6970
6940 LET Y1 = ASC("L")
6950 LET X0 = 1
6960 GO TO 7000
6970 IF X0 <= M THEN 7060
6980 LET Y1 = ASC("R")
6990 LET X0 = M
```

```
7000 REM OUTLIER IN 1 OR 2 DIRECTIONS?

7010 IF Y0 <= 9 THEN 7060
7020 LET Y1 = ASC("*")

7030 REM ALWAYS FAVOR THE MORE EXTREME VALUE
7040 REM   DONT OVERWRITE OUTLIERS
7050 REM >>VERY ASCII-DEPENDENT CODE HERE

7060 IF P(X0) = ASC("*") THEN 7200
7070 IF Y1 = ASC("*") THEN 7110
7080 IF P(X0) = 92 THEN 7110
7090 IF P(X0) > ASC("9") THEN 7150
7100 IF P(X0) >= Y1 THEN 7200
7110 LET P(X0) = Y1
7120 IF P6 >= X0 THEN 7200
7130 LET P6 = X0
7140 GO TO 7200

7150 REM   EITHER  L,R,M,OR P IN Y(X0) ALREADY

7160 IF Y1 <= ASC("9") THEN 7200
7170 IF Y1 = P(X0) THEN 7200
7180 LET Y1 = ASC("*")
7190 GO TO 7110
7200 IF K > 1 THEN 6720

7210 REM   PRINT THE LINE

7220 PRINT TAB(M0 + 4);
7230 FOR I = 1 TO P6
7240    PRINT CHR$(P(I));
7250 NEXT I
7260 PRINT
7270 IF K > 1 THEN 6240

7280 REM   IF MORE TO PLOT, GO DO IT. ELSE SORT ON X() AND RETURN

7290 GOSUB 1200
7300 RETURN
```

```
      SUBROUTINE PLOT(Y, X, N, WY, WX, LINSET, CHRSET, XMIN, XMAX,
     1 YMIN, YMAX, ERR)
C
C PLOT THE  N  ORDERED PAIRS (X(I), Y(I)) USING A CONDENSED PLOT.
C CONDENSED PLOTTING USES THE PLOTTING SYMBOL TO INDICATE THE FINE
C DETAIL OF VERTICAL SPACING.  AS A RESULT, MORE PRECISION CAN BE
C CONVEYED IN FEWER LINES.  MULTIPLE POINTS FALLING AT THE SAME
C PLOT POSITION ARE NOT INDICATED, HOWEVER -- THE MOST EXTREME
C (IN Y) POINT WILL BE SELECTED FOR DISPLAY.
C  X() AND Y() ARE NOT MODIFIED BY THE PROGRAM.  WORK IS DONE USING
C THE WORK ARRAYS  WY() AND WX()  SUPPLIED BY THE CALLING PROGRAM.
C  THE DETAILS OF PLOT FORMAT ARE DETERMINED BY THE PARAMETERS IN THE
C CALLING SEQUENCE.  LINSET SPECIFIES THE MAXIMUM NUMBER OF LINES TO
C BE USED.  CHRSET SPECIFIES HOW MANY DIFFERENT CODES CAN BE USED ON
C EACH LINE.  IF EITHER OF THESE IS ZERO, THE PROGRAM DEFAULTS TO
C 10 LINES AND 10 CHARACTER CODES (0 THRU 9).
C XMIN AND XMAX SPECIFY THE RANGE OF X-VALUES TO BE PLOTTED.
C YMIN AND YMAX SPECIFY THE RANGE OF Y-VALUES TO BE PLOTTED.
C FOR EITHER PAIR, IF THEY ARE SET EQUAL BY THE CALLING PROGRAM,
C THE PROGRAM DEFAULTS TO USING THE ADJACENT VALUES IN EACH DIMENSION.
C THIS OPTION IS ALMOST ALWAYS PREFERRED FOR EXPLORATORY PLOTS.
C
      INTEGER N, LINSET, CHRSET, ERR
      REAL Y(N), X(N), WY(N), WX(N), XMIN, XMAX, YMIN, YMAX
C
      COMMON /CHRBUF/ P, PMAX, PMIN, OUTPTR, MAXPTR, OUNIT
      INTEGER P(130), PMAX, PMIN, OUTPTR, MAXPTR, OUNIT
C
C FUNCTIONS
C
      INTEGER INTFN, FLOOR, WDTHOF
C
C CALLS SUBROUTINES NPOSW, PRINT, PSORT, PUTCHR, PUTNUM, YINFO
C
C LOCAL VARIABLES
C
      INTEGER CHL, CHM, CHP, CHR, CHO, CH9, CHPLUS, CHMIN
      INTEGER CHRPAR, CHSTAR
      INTEGER LINES, CHRS, MAXL, XPOSNS, IADJL, IADJH, NN, LFTPSN
      INTEGER LNSFRZ, LINENO, PTR, NWID, PROOM, OCHAR, OPOS, YCHAR
      INTEGER OPOSX, LNFLOR, LABEL, I
C
      REAL HH, HL, MED, STEP, TOP, BOTTOM, LEFT, RIGHT
      REAL ADJXL, ADJXH, ADJYL, ADJYH, XFRACT, XUNIT, XNPW, YFRACT
      REAL YUNIT, YNPW, YLABEL, XVAL, SYVAL, NICNOS(9)
      LOGICAL NEGNOW, MARKZS
```

```
C
          DATA CHL, CHM, CHP, CHR, CHO, CH9/12, 13, 16, 18, 27, 36/
          DATA CHPLUS,CHMIN,CHSTAR,CHRPAR/39,40,41,44/
          DATA NN, NICNOS(1), NICNOS(2), NICNOS(3) /9, 1.0, 1.5, 2.0/
          DATA NICNOS(4), NICNOS(5), NICNOS(6) /2.5, 3.0, 4.0/
          DATA NICNOS(7), NICNOS(8), NICNOS(9) /5.0, 7.0, 10.0/
          DATA MARKZS /.FALSE./
C
C
          IF ( N .GE. 5) GO TO 10
          ERR = 41
          GO TO 999
   10     LFTPSN = PMIN + 6
          LINES = 10
          CHRS = 10
          IF(LINSET .EQ. 0  .OR.  CHRSET .EQ. 0) GO TO 30
          LINES = LINSET
          CHRS = CHRSET
          ERR = 42
          IF(LINES .LT. 5  .OR.  LINES .GT. 40) GO TO 999
          IF(CHRS .LT. 1 .OR.  CHRS .GT. 10) GO TO 999
          ERR = 0
C
C  SET UP SCALES AND PLCT BOUNDARY INFORMATION
C
   30     LFTPSN = PMIN + 6
          PROOM = PMAX - LFTPSN + 1
          DO 40 I = 1, N
            WX(I) = X(I)
   40     CONTINUE
          IF(XMIN .GE. XMAX) GO TO 45
          CALL YINFO(WX, N, MED, HL, HH, ADJXL, ADJXH, IADJL, IADJH,
     1      STEP, ERR)
          IF(ERR .NE. 0) GO TO 999
          IF(ADJXL .LT. ADJXH) GO TO 50
C
C  IF X-ADJACENT VALUES EQUAL, TRY USING THE EXTREMES
C
   45     ADJXL = WX(1)
          ADJXH = WX(N)
          ERR = 44
          IF(ADJXL .GE. ADJXH) GO TO 999
          ERR = 0
   50     CALL NPOSW(ADJXH, ADJXL, NICNOS, NN, PROOM, .FALSE., XPOSNS,
     1    XFRACT, XUNIT, XNPW, ERR)
```

```
C
C    SCALE Y --SORT (X, Y) PAIRED ON Y
C
         DO 60 I = 1, N
            WX(I) = X(I)
            WY(I) = Y(I)
   60    CONTINUE
         CALL PSORT(WY, WX, N, ERR)
         IF(YMIN .GE. YMAX) GO TO 65
         CALL YINFO(WY, N, MED, HL, HH, ADJYL, ADJYH, IADJL, IADJH,
      1    STEP, ERR)
         IF(ERR .NE. 0) GO TO 999
         GO TO 68
   65    ADJYL = WY(1)
         ADJYH = WY(N)
         ERR = 45
         IF(ADJYL .GE. ADJYH) GO TO 999
         ERR = 0
   68    MAXL = LINES
         CALL NPOSW(ADJYH, ADJYL, NICNOS, NN, MAXL, .TRUE., LINES,
      1    YFRACT, YUNIT, YNPW, ERR)
         IF(ERR .NE. 0) GO TO 999
         IF (YFRACT .NE. 10.0) GO TO 70
         YFRACT = 1.0
         YUNIT = YUNIT*10.0
C
C    FIND NICE PLOT EDGES -- ROUND AWAY FROM CENTER OF PLOT
C
   70    LNSFRZ = -FLOOR(-ADJYH/YNPW)
         TOP = FLOAT(LNSFRZ) * YNPW
         BOTTOM = FLOAT(FLOOR(ADJYL/YNPW)) * YNPW
         LEFT = FLOAT(FLOOR(ADJXL/XNPW)) * XNPW
         RIGHT = FLOAT(-FLOOR(-ADJXH/XNPW)) * XNPW
C
C    PRINT SCRAWL
C
         WRITE(OUNIT, 9070) LINES, CHRS
 9070    FORMAT(15X, I3, 7H LINE, , I3, 15H CHARACTER PLOT)
         WRITE(OUNIT, 9080)BOTTOM, TOP, YNPW, LEFT, RIGHT, XNPW
 9080    FORMAT(15X, 8H Y FROM , F12.6, 4H TO , F12.6, 7H  STEP , F12.6/
      1    15X, 8H X FROM , F12.6, 4H TO , F12.6, 7H  STEP , F12.6//)
C    INITIALIZE FOR PLOTTING--ONE LINE TOO HIGH
C    LNSFRZ COUNTS # LINES AWAY FROM ZERO--+00 AND -00 ARE 0 LINES AWAY.
C
         YLABEL = FLOAT(LNSFRZ) * YFRACT
         LNFLOR = LNSFRZ
         NEGNOW = .FALSE.
         IF(TOP .GT. 0.0) GO TO 80
         LNSFRZ = LNSFRZ + 1
         NEGNOW = .TRUE.
   80    LINENO = 0
         PTR = N+1
```

```
C
C   START A NEW LINE OF THE PLOT
C
    90    LNFLOR = LNFLOR - 1
          LINENO = LINENO + 1
          OPOS = PMIN
          IF(LNSFRZ .GT. 0 .OR. NEGNOW) GO TO 95
C
C   JUST WENT NEGATIVE
C
          NEGNOW = .TRUE.
          GO TO 97
    95    LNSFRZ = LNSFRZ - 1
          YLABEL = YLABEL - YFRACT
    97    CONTINUE
C
C   PRINT THE LINE LABEL
C
          IF(.NOT. NEGNOW) CALL PUTCHR(OPOS, CHPLUS, ERR)
          IF(NEGNOW) CALL PUTCHR(OPOS, CHMIN, ERR)
          IF(YLABEL .NE. 0.0) GO TO 120
          OPOS = PMIN + 3
          CALL PUTCHR(OPOS, CHO, ERR)
          CALL PUTCHR(O, CHO, ERR)
          IF((CHRS .GT. 1) .AND. ((LINES-LINENO) .GE. 3)) MARKZS = .TRUE.
          OPOS = PMIN + 5
          CALL PUTCHR(OPOS, CHRPAR, ERR)
          IF(.NOT. MARKZS) GO TO 111
          DO 100 I = 1, 5
             IF(.NOT. NEGNOW) CALL PUTCHR(O, CHMIN, ERR)
             IF(NEGNOW) CALL PUTCHR(O, CHMIN, ERR)
    100   CONTINUE
          OPOSX = PMAX - 5
          DO 110 OPOS = OPOSX, PMAX
             IF(.NOT. NEGNOW) CALL PUTCHR(O, CHMIN, ERR)
             IF(NEGNOW) CALL PUTCHR(O, CHMIN, ERR)
    110   CONTINUE
    111   CONTINUE
          GO TO 125
C
C    PRINT NON-ZERO LABEL
C
    120   LABEL = INTFN(10.0 * ABS(YLABEL), ERR)
          IF( ERR .NE. 0) GO TO 999
          NWID = WDTHOF(LABEL)
          OPOS = PMIN + 5 - NWID
          CALL PUTNUM(OPOS, LABEL, NWID, ERR)
          IF (ERR .NE. 0) GO TO 999
C
C   GET NEXT DATA POINT
C
    125   PTR = PTR - 1
          IF(PTR .LE. 0) GO TO 135
          XVAL = WX(PTR)
          SYVAL = WY(PTR)/YNPW
```

```
           IF(INTFN(SYVAL, ERR) .GT. LNFLOR) GO TO 140
           IF(INTFN(SYVAL, ERR).EQ.LNFLOR .AND. SYVAL .GE. 0.0) GO TO 140
C
C  TIME TO START NEXT LINE
C  IF THIS IS THE LAST LINE, PRINT IT ANYWAY AND USE "M" FOR LOW NO.
C
  130    IF(LINENO .EQ. LINES) GO TO 140
C
C  BACK UP THE POINTER
C
         PTR = PTR + 1
C
C  WRAP UP LINE
C
  135    IF(ERR .NE. 0) GO TO 999
         CALL PRINT
C
C  AND START A NEW LINE
C
         GO TO 90
C
C  GET Y-CHARACTER
C
  140    YCHAR = IFIX(ABS(SYVAL - FLOAT(LNSFRZ)) * FLOAT(CHRS))
         OCHAR = CHO + YCHAR
         IF(OCHAR .GE. CHO .AND. OCHAR .LE. CH9) GO TO 145
         OCHAR = CHP
         IF(LINENO .EQ. LINES) OCHAR = CHM
C
C  GET X-POSITION
C
  145    OPOS = PMIN + 5 + INTFN((XVAL - LEFT)/XNPW, ERR) + 1
         IF (XVAL .GE. LEFT) GO TO 150
         OPOS = PMIN + 6
         IF(OCHAR .LT. CHO .OR. OCHAR .GT. CH9) GO TO 147
         OCHAR = CHL
         GO TO 160
  147    OCHAR = CHSTAR
         GO TO 160
  150    IF(XVAL .LE. RIGHT) GO TO 160
         OPOS = PMAX
         IF(OCHAR .LT. CHO .OR. OCHAR .GT. CH9) GO TO 157
         OCHAR = CHR
         GO TO 160
  157    OCHAR = CHSTAR
  160    CONTINUE
         CALL PUTCHR(OPOS, OCHAR, ERR)
         IF(ERR .NE. 0) GO TO 999
         IF(PTR .GT. 1) GO TO 125
         CALL PRINT
  999    RETURN
         END
```

Chapter 5 _____

Resistant Line

In Chapter 4 we focused our attention on flexible techniques for plotting a

response
factor

response, y, against a *factor, x.* When the pattern of a plot suggests that the value of y depends on the value of x, we often try to summarize this dependence in terms of the simplest possible description—namely, a straight line. We can represent any straight line with the equation

$$y = a + bx$$

just by choosing values for a and b. Once we have a and b, every pair of numbers (x, y) that satisfies the relationship $y = a + bx$ will lie on a straight line when plotted. In order to summarize any particular x-y data, we need numerical values for a and b that will make a line pass close to the data. This chapter shows one way to find these values.

5.1 Slope and Intercept

slope

The numbers represented by a and b in the equation of a line have specific meanings. The *slope* of the line, b, tells us how tilted the line is; more precisely,

121

intercept

it tells us the change in y associated with a one-unit increase in x. The *intercept*, a, is the height (level) of the line when x equals zero—that is, the value of y where the line crosses the y-axis.

The slope and intercept of any straight line can be found from any two points on the line. For example, we can choose a point on the left with a low x-value—labeled (x_L, y_L) in Exhibit 5–1—and a point on the right with a high x-value—labeled (x_R, y_R). The slope, b, is defined as the change in y divided by the corresponding change in x. Writing this quotient precisely with our two points gives

$$b = \frac{y \text{ change}}{x \text{ change}} = \frac{y_R - y_L}{x_R - x_L}.$$

One common way to describe the slope is "change in y *per* change in x." For example, the statement "sales have grown by 2500 dollars per year" specifies a slope.

When we know b, we can find the intercept by using either of these points and specifying that the line must pass through it. For example, $y_L = a + bx_L$, where we already know b. Solving for a, we get

$$a = y_L - bx_L.$$

Exhibit 5–1 Finding the Slope and Intercept of the Line $y = a + bx$

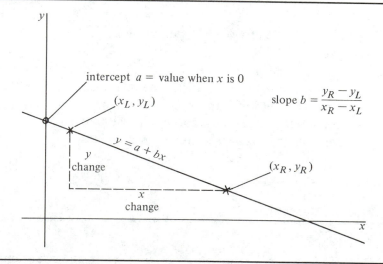

Note: In this example y_R is smaller than y_L so $y_R - y_L$ is negative and the slope, b, is also negative.

We can equally well get

$$a = y_R - bx_R.$$

Exhibit 5–1 shows the geometry behind these calculations.

5.2   Summary Points

When we deal with a line itself, it doesn't matter which two points we use to calculate a and b because every point we consider is exactly on the line. However, we can't expect real data to line up perfectly. While many points may be near a line, few will lie exactly on it. Many different lines could pass close enough to the data to be reasonable summaries. Consequently, we can't just pick any two points from the data and expect to find a good line. Instead we want to find points that summarize the data well so that the line they determine will be close to the data.

To get an estimate of the slope, we need to pick a typical x-value near each end of the range of x-values but not so near as to risk being an extraordinary x-value. We do this by dividing the data into three portions or regions—points with low x-values (on the left), points with middle x-values, and points with high x-values (on the right)—with roughly a third of the points in each portion. Exhibit 5–2 illustrates this partitioning. If we can't put exactly the same number of points into each portion because $n/3$ leaves a remainder, we still allocate the points symmetrically. A single "extra" point goes into the middle portion; when two "extra" points remain, one goes into each outer portion. Whenever several data points have the same x-value, they must go into the same portion. Such ties may make it more difficult to come close to equal allocation. When we work by hand, we can usually use our judgment to resolve the problem of equal allocation. Precise rules to handle all situations may be found in the programs at the end of this chapter.

Within each portion (or third) of the data, we forget about the pairing between the x-value and the y-value in x-y data and summarize the x-values and the y-values separately. In each portion, we first treat the x-values as a batch (and ignore y) and find their median. We then treat the corresponding y-values as a batch and find *their* median. Thus, we obtain an (x, y) pair of medians in each of the three portions. The points that these median pairs specify need not be original data points, but they may be. Nothing forces the median x-value and the median y-value to come from the same data point, even though the assignment of y-values to portions is determined entirely by

Exhibit 5–2 Dividing a Plot into Thirds and Finding Summary Points

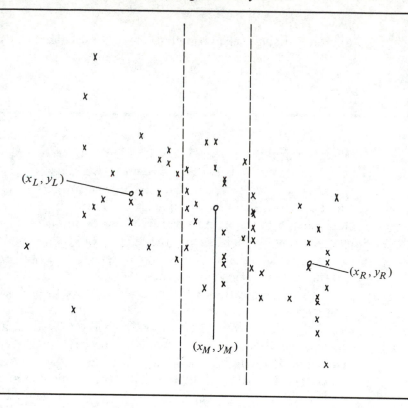

the *x*-values. For example, when the data points lie very close to a line with a steep slope, the *y*-value order of the points will be the same as their *x*-value order, and the median *x*-value and median *y*-value will come from the same data point.

Because these points are chosen from the middle of each third of the data, they summarize the behavior of the batch in each region. Accordingly, they are called *summary points*. If we label the thirds as left (*L*), middle (*M*), and right (*R*) according to the order of the *x*-values, the three summary points can be denoted by

summary points

$$(x_L, y_L)$$

$$(x_M, y_M)$$

$$(x_R, y_R).$$

SONESTA HOTELS

Exhibit 5–2 shows the three summary points for one data batch. As we will see, using the median in finding the summary points makes the line *resistant* to stray values in the y- *or* x-coordinate of the data points.

5.3   Finding the Slope and the Intercept

Once we have found the summary points, we can easily calculate the values of a and b. For the slope, b, we return to its definition and divide the change in y between the outer summary points, $y_R - y_L$, by the change in x between these same points, $x_R - x_L$. Thus we find

$$b = \frac{y_R - y_L}{x_R - x_L}.$$

The intercept, a, should be adjusted to make the line pass, as nearly as possible, through the middle of the data. We could make it pass through the middle summary point by computing the needed adjustment from that point:

$$a = y_M - bx_M.$$

However, rather than allow the middle summary point alone to determine the intercept, we use all three summary points and average the three intercept estimates:

$$a_L = y_L - bx_L$$
$$a_M = y_M - bx_M$$
$$a_R = y_R - bx_R$$

and hence

$$a = (\tfrac{1}{3})(a_L + a_M + a_R) = (\tfrac{1}{3})[(y_L + y_M + y_R) - b(x_L + x_M + x_R)].$$

5.4   Residuals

residuals

model

fit

A fundamental step in most data analysis and in all exploratory analysis is the computation and examination of *residuals*. While we usually begin to examine data with some elementary displays such as those presented in Chapters 1 through 4, most analyses propose a simple structure or *model* to begin describing the patterns in the data. Such models differ widely in structure and purpose, but all attempt to fit the data closely. We therefore refer to any such description of the data as a *fit*. The residuals are, then, the differences at each point between the observed data value and the fitted value:

$$residual = data - fit.$$

The resistant line provides one way to find a simple fit, and its residuals, r, are found for each data value, (x_i, y_i), as

$$r_i = y_i - (a + bx_i).$$

A pessimist might view residuals as the failure of a fit to describe the data accurately. He might even speak of them as "errors," although a perfect fit, which leaves all residuals equal to zero, would arouse suspicion. An optimist sees in residuals details of the data's behavior previously hidden beneath the dominant patterns of the fit. Both points of view are correct. The best fits leave small residuals, and systematically large residuals may indicate a poorly chosen model. Nevertheless, even a good fit may do nothing more than describe the obvious—for example, prices increased during the 1970s; the population of the United States grew during the same period—and leave behind the interesting patterns—for example, the Vietnam war affected the U.S. economy; the birthrate dropped sharply.

Any method of fitting models must determine how much each point can be allowed to influence the fit. Many statistical procedures try to keep the fit close to *every* data point. If the data include an outlier, these procedures may permit it to have an undue influence on the fit. As always in exploratory data analysis, we try to prevent outliers from distorting the analysis. Using medians in fitting lines to data provides resistance to outliers, and thus the

resistant line

line-fitting technique of this chapter is called the *resistant line*.

5.5   Polishing the Fit

Resistance to outliers has one price. The values found at first for the intercept, *a*, and the slope, *b*, are often not the most appropriate ones. A good way to check the values we have found is to calculate the residuals, treat the points

$$(x, residual) = (x_i, y_i - (a + bx_i)),$$

as *x-y* data, and find summary points as before. If the slope, *b'*, between the outer summary points is zero (or very close to zero), we are done. If not, we can adjust the original slope by adding the residual slope *b'* to it. We will, of course, want to compute the new residuals to see whether *their* slope is now close enough to zero.

Sometimes we will have overcorrected, and the new residuals will tilt the other way. When we have two slopes, one too small (residuals have a positive slope) and one too large (residuals have a negative slope), we know that the correct slope lies between them. We can often improve the slope estimate very efficiently by using the correction formula

$$b_{new} = b_2 - b'_2 \, [(b_2 - b_1)/(b'_2 - b'_1)].$$

Here b_1 and b_2 are the two slope estimates, and b'_1 and b'_2 are the slopes of the residuals when b_1 and b_2 were tried. The example in the next section illustrates this process and shows how still more corrections can be made if needed.

5.6   Example: Breast Cancer Mortality versus Temperature

In a 1965 report, Lea discussed the relationship between mean annual temperature and the mortality rate for a type of breast cancer in women. The data, pertaining to certain regions of Great Britain, Norway, and Sweden, are listed in Exhibit 5–3 and are plotted in Exhibit 5–4.

In this example, $n = 16$ and $n/3 = 5\frac{1}{3}$. To keep the thirds symmetric, we want to allocate the spare data value to the middle third in order to have 5 points in the left third, 6 in the middle third, and 5 in the right third; because

Exhibit 5–3   Mean Annual Temperature (in °F) and Mortality Index for Neoplasms of the Female Breast

Mean Annual Temperature	Mortality Index
51.3	102.5
49.9	104.5
50.0	100.4
49.2	95.9
48.5	87.0
47.8	95.0
47.3	88.6
45.1	89.2
46.3	78.9
42.1	84.6
44.2	81.7
43.5	72.2
42.3	65.1
40.2	68.1
31.8	67.3
34.0	52.5

Source: Data from A.J. Lea, "New Observations on Distribution of Neoplasms of Female Breast in Certain European Countries," *British Medical Journal* 1 (1965):488–490. Reprinted by permission.

no two *x*-values are the same, we can do exactly this. Ordering the (x, y) points from lowest to highest *x*-value and separating the thirds, we obtain the first two columns of Exhibit 5–5. It is now a straightforward matter to find the *x*- and *y*-components of the summary points:

Third	Median x	Median y
L	40.2	67.3
M	45.7	85.15
R	49.9	100.4

(In finding the summary values, we are reminded that the value or values that determine median *x* and those that determine median *y* need not come from the same data points.) Now the initial value of *b* is

$$b = \frac{y_R - y_L}{x_R - x_L} = \frac{100.4 - 67.3}{49.9 - 40.2} = 3.412,$$

Exhibit 5–4   Mortality Index versus Mean Annual Temperature for the Breast Cancer Data of Exhibit 5–3

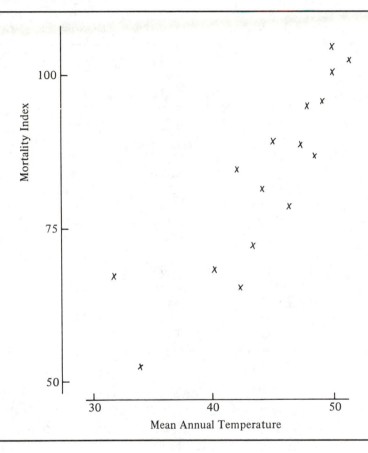

and that of *a* is

$$a = \tfrac{1}{3}[(y_L + y_M + y_R) - b(x_L + x_M + x_R)]$$

$$= \tfrac{1}{3}[(252.85) - 3.412 \times (135.8)] = -70.17.$$

Thus the initial fitted line is

$$y = -70.17 + 3.412x,$$

where *y* = *mortality index* and *x* = *mean annual temperature*. Now, at each

Exhibit 5–5   Calculating Resistant Line for Breast Cancer Mortality Data of Exhibit 5–3

(x) Temperature	(y) Mortality	First Residual	Fourth Residual	Final Residual
31.8	**67.3**	28.97	45.57	21.59
34.0	52.5	**6.66**	**24.41**	0.43
40.2	68.1	1.11	22.09	−1.89
42.1	84.6	11.12	33.10	9.12
42.3	65.1	−9.06	13.02	−10.96
43.5	72.2	−6.05	16.66	−7.32
44.2	**81.7**	**1.06**	**24.13**	0.15
45.1	89.2	5.49	29.03	5.05
46.3	78.9	−8.91	15.26	−8.72
47.3	**88.6**	**−2.62**	**22.07**	−1.91
47.8	95.0	2.08	27.03	3.05
48.5	87.0	−8.31	17.00	−6.98
49.2	95.9	**−1.80**	23.88	−0.10
49.9	104.5	4.41	30.46	6.48
50.0	**100.4**	−0.03	26.07	2.09
51.3	102.5	−2.37	**24.41**	0.43

first residuals   point we subtract the fitted value found by this line from the observed *y*-value, according to $y_i - (a + bx_i)$. The subtraction yields the column of *first residuals* in Exhibit 5–5 and completes the first iteration in the process of fitting a resistant line to this set of data.

We can now compute the slope of these residuals. We find the median of the first residuals in each portion and, from them, correction summary points,

$$(40.2, 6.66)$$

$$(45.7, -0.78)$$

$$(49.9, -1.80),$$

and the slope of the residuals,

$$b' = \frac{-1.80 - 6.66}{49.9 - 40.2} = -0.872.$$

The second slope estimate is then

$$b_2 = 3.412 - 0.872 = 2.540.$$

The residuals from the line with this slope and the original intercept are the "second residuals." *Their* slope, b'_2, is found in the same way. Here it is 0.624. We could adjust the intercept as well, but it is easier to wait until we have a satisfactory slope estimate.

We now have two slope estimates, 3.412, and 2.540, which leave residual slopes with opposite signs: -0.872 and 0.624. These are all we need to apply the second correction formula. We compute a new slope estimate as

$$b_3 = 2.540 - 0.624[(2.540 - 3.412)/(0.624 - (-0.872))] = 2.904.$$

We then compute the residuals from the line with slope b_3 and find their slope. In this example, $b'_3 = -0.024$—much closer to zero than the previous residual slopes.

Although a residual slope of -0.024 is small enough for most purposes, we will try one more correction step. Because the final slope must lie between a slope estimate that is too low (with positively sloped residuals) and one that is too high (with negatively sloped residuals), we use the current best guesses for these two estimates. Our latest estimate has negatively sloped residuals ($b'_3 = -0.024$), so we use it and its residual slope in place of our former high slope estimate, 3.412. This yields

$$b_4 = 2.904 - (-0.024)[(2.904 - 2.540)/(-0.024 - 0.624)] = 2.890.$$

The residuals from the line with slope b_4 and the original intercept are in the column of "fourth residuals" in Exhibit 5–5. They have slope 0.0, so no further adjustment is possible. Exhibit 5–6 summarizes these steps.

We can now compute the intercept using the summary points of the fourth residuals. We find

$$a_4 = \tfrac{1}{3}(24.41 + 23.10 + 24.41) = 23.98.$$

Thus the final fit is

$$y = (-70.17 + 2.890x) + 23.98 \text{ or } y = -46.19 + 2.890x.$$

We interpret this line as saying that mortality from this type of breast cancer increases with increasing mean annual temperature at the rate of about 2.9

Exhibit 5–6 The Resistant Line Iterated to "Convergence" for the Breast Cancer Mortality Data of Exhibit 5–3

Slope 1: 3.412
Slope 2: 2.540
Slope 3: 2.904
Slope 4: 2.890
Fitted line: $y = -46.2 + 2.890x$

---

mortality index units per *degree Fahrenheit.* The intercept of the final line has no simple interpretation here except perhaps that if this trend held for colder climates, the breast cancer mortality index would approach zero where the mean annual temperature was 16.0° (because $2.890 \times 16.0 = 46.2$).

When we work by hand, we will usually stop with the second or third slope estimate. When we can use a computer, a few more steps will often yield the slope estimate with zero residual slope.

A few hints make the calculations easier: To use the second correction formula, we need two slopes, one too high and one too low. If the slope of the second residuals is not opposite in sign to the slope of the first residuals, we must try larger corrections to the first slope estimate until the second residuals tilt the other way. (This happens in a later example; see Exhibit 5–15.)

When we have two slope estimates and solve for the next estimate with the formula

$$b_{new} = b_2 - b_2'[(b_2 - b_1)/(b_2' - b_1')],$$

it does not matter which slope is used for b_1 and which for b_2. However, it is usually best to choose as b_2 the slope estimate with smaller residual slope.

We can save computing in two ways. First, we need not find the middle-third residuals until we have settled on a final slope. Second, we can replace b' by the difference between the right and left median residuals. A little algebra shows that the divisor $(x_R - x_L)$ in the slope calculations cancels out the formula for b_{new}, so we can avoid dividing by it.

We always examine the residuals by displaying them in a stem-and-leaf display and plotting them against x. Exhibits 5–7 and 5–8 show these displays of the residuals, and Exhibit 5–5 lists the final residuals for comparison with earlier steps. The most noticeable feature in the plot of the residuals is the high point at the left. We already noticed this deviant point in Exhibit 5–4, and the residuals are now telling us that it did not twist the resistant line. A closer look at Exhibit 5–8, along with an examination of the sign pattern of the

Exhibit 5–7   Final Residuals from Exhibit 5–5

```
                    STEM-AND-LEAF DISPLAY
                    UNIT = 1
                    1  2  REPRESENTS 12.


                    1    -1*    0
                    2    -0·    8
                    4     S     76
                    4     F
                    4     T
                    7    -0*    110
                    9    +0*    000
                    6     T     23
                    4     F     5
                    3     S     6
                    2    +0·    9

                    HI :    21,
```

Exhibit 5–8   Plot of Final Residuals against Mean Annual Temperature

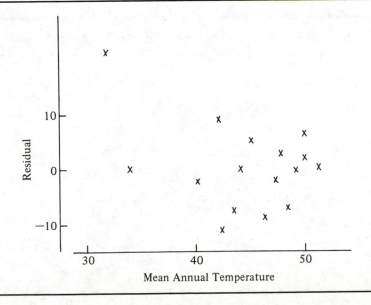

residuals in Exhibit 5–5, reveals an unusual pattern—four parallel diagonal bands of points plus two points at very low *x*-values and one at a high *x*-value. Although no explanation for this pattern is evident, it may deserve further attention.

5.7   Outliers

In previous chapters, outliers were principally identified as data values that are extraordinary on a single variable. By separating the data values into a fit and a set of residuals, we are able to think about outliers in greater detail.

When we consider *y*-versus-*x* relationships, we must beware of points that are extraordinary in *y*, in *x*, or in both simultaneously. Luckily, the resistant line protects our analysis from most of the effects of such points. Often the more interesting data points are those with extreme *residuals*. These points are not well described by the fit and should therefore receive further attention. They need not be outliers in either *x* or *y* alone. Exhibit 5–9 shows a plot of age-adjusted mortality rate versus median education for the same 60 United States SMSAs considered in other examples. (See Exhibit 1–4 for the data.) There is a clear trend: Higher median education is associated with lower mortality rates. However, two SMSAs stand out as having a much lower mortality rate than other SMSAs with similar median education levels. These

Exhibit 5–9   Age-Adjusted Mortality versus Median Education for 60 U.S. SMSAs

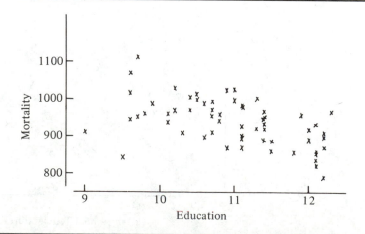

two are York and Lancaster, Pennsylvania, which both contain many Amish, who traditionally have expected a minimum amount of formal education of their children. While these two SMSAs do have the lowest median education levels of the 60 SMSAs reported, the median education levels are certainly not extraordinary in themselves. What is remarkable is the large deviation of these values from the general trend—a deviation that would show up as a large residual from a resistant line.

Alternatively, it is possible for points extraordinary in x and y to have small residuals. This is likely when the x-value and y-value are naturally extreme but not erroneous—that is, when the point is well described by the fit but lies far from most of the data.

Data values with outlying residuals should be treated in much the same way as simple outliers. We check for errors, and, if we cannot correct them, we consider omitting these data values. If we believe the numbers to be correct, we look for possible additional information to help explain their nonconformity. This search, in particular, is often well worth the effort because explainable outliers often yield much valuable insight.

5.8   Straightening Plots by Re-expression

A straight line is a desirable summary for an x-y relationship because of its simplicity of form and of interpretation. However, the relationship between y and x need not be linear. We can examine the shape of the relationship with an x-y plot and look for more detailed information by plotting the residuals from a resistant line against x. If either the original or residual plot shows a bend and if the y-versus-x plot shows a generally consistent trend either up or down rather than a cup shape, we may be able to straighten the y-versus-x relationship by re-expressing one or both variables. Once again we will limit our choice of re-expressions to the ladder of powers (see Section 2.4); and, as before, we find that the ordering of powers also orders their effects.

We can get an idea of how straight the relationship between x and y is by using the three summary points (Section 5.2). We approximate the slope in *half-slopes*   each half of the data by computing the left and right *half-slopes*,

$$b_L = \frac{y_M - y_L}{x_M - x_L} \quad \text{and} \quad b_R = \frac{y_R - y_M}{x_R - x_M},$$

half-slope   and then we find the *half-slope ratio*, b_R/b_L. If the half-slopes are equal, then *ratio*   the x-y relationship is straight and the half-slope ratio is 1. If the half-slope

Exhibit 5–10   Patterns in *x-y* Relationships Point the Direction of Re-expressions on the Ladder of Powers

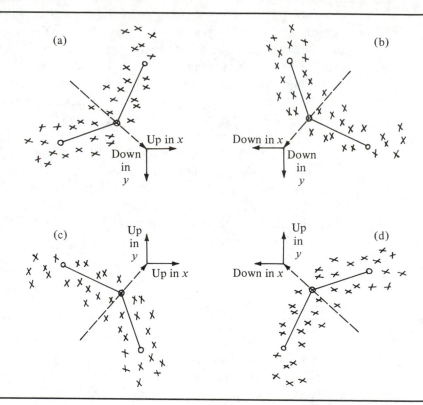

ratio is not close to 1, then re-expressing *x* or *y* or both may help. If the half-slope ratio is negative, the half-slopes have different signs, and re-expression will not help.

If the half-slopes are not equal, the plotted line segment joining the left and middle summary points will meet the line segment joining the middle and right summary points at an angle, as shown in Exhibit 5–10. We can think of this angle as forming an arrowhead that points toward re-expressions on the ladder of powers that might make the relationship straighter. To determine how we might re-express *y*, we ask whether the arrow points more upward—toward higher *y*-values—or more downward—toward lower *y*-values. (The half-slopes must have the same sign if re-expression is to help; so the arrow cannot point directly to the right or to the left.) To determine how we might re-express *x*, we ask whether the arrow points more to the right—toward higher *x*-values—or more to the left—toward lower *x*-values. Exhibit 5–10 shows the four possible patterns.

Thus, the rule for selecting a re-expression to straighten a plot is that we consider moving the expression of y or x in the direction the arrow points. That is, if the arrow points down, toward lower y, we might try re-expressions of y lower on the ladder of powers. Recall that raw data is the 1 power; so, moving down the ladder, we would try $\sqrt{y}$ ($\frac{1}{2}$ power), $\log(y)$ (0 power), $-1/\sqrt{y}$ ($-\frac{1}{2}$ power), and so on. If the arrow points to the right, toward higher x, we might try re-expressions of x higher on the ladder of powers, such as x^2 or x^3.

As we saw when we re-expressed data to improve symmetry, the ladder of powers orders re-expressions according to the strength of their effect. Thus, if the half-slope ratio is well above 1 and the bend in the plot suggests moving down the ladder of powers in y, $\sqrt{y}$ will probably be straighter against x. If $\sqrt{y}$ still shows a bend pointing toward lower y-values, then $\log(y)$ is likely to be better. Of course, if we move far enough down the ladder of powers, the half-slope ratio will eventually fall below 1, and the bend in the plot will point the other way. Thus we can systematically seek a re-expression by examining the half-slope ratio and letting it guide changes to stronger or less strong re-expressions.

A little thought will reveal how re-expressing can straighten an x-y relationship and why this mnemonic rule works. If the half-slopes point down and to the right, as in part (a) of Exhibit 5–10, the higher y-values need to be pulled together more to straighten the relationship. This is what re-expressions lower than 1 on the ladder of powers do. For example, 0, 25, 100, and 225 are made equally spaced by a square-root re-expression, and 1, 10, 100, and 1000 are made equally spaced by a log re-expression. If larger y-values grow more rapidly than smaller y-values, re-expressing y by square roots or logs (or some lower power) is likely to slow their growth and make the relationship straighter.

An alternative interpretation of the "down and to the right" pattern is to stretch out the higher x-values so that they grow as rapidly as their corresponding y-values. Re-expressions above 1 on the ladder of powers do this. For example, 0, 5, 10, and 15 are stretched to 0, 25, 100, and 225 by squaring and to 0, 125, 1000, and 3375 by cubing.

Thus, re-expressions alter the shape of data by stretching or shrinking the larger values differently from the smaller ones. Consequently, data batches in which the larger values are many times larger than the smaller ones will be more affected by re-expressing than will batches in which the largest and smallest values are of about the same magnitude. Re-expressing data that range from 10.3 to 13.8 is pointless, but data stretching from 3 to 3000 will respond to even a small move along the ladder of powers.

The pair of half-slope lines meeting at the middle summary point will, of course, suggest re-expressions for both x and y. We may choose to re-express either y or x or both. Often the nature of the data will lead us to prefer re-expressing one or the other. Sometimes a particular re-expression for *x* or *y* will be suggested by the units in which the data are measured or by some other aspect of the data, but if re-expressing one of *x* and *y* does not straighten the relationship sufficiently, we might try re-expressing the other. If either *x* or *y* covers a much greater range of magnitude than the other, it will be more affected by re-expression, so we might try to re-express it first and use the other to "fine tune" the result. Finally, we often prefer to re-express *y*, simply because we think of *x* as the circumstance or the base from which to predict or describe *y*, and thus we prefer to have *x* in its original units.

When we work on a computer, we usually will not mind re-expressing all of the *x*- or *y*-values, computing a new half-slope ratio, and drawing a new plot. When we work by hand (or when getting the results from the computer takes too long or costs too much), we can learn almost as much from the three summary points alone. The summary points of re-expressed data can be found by re-expressing the appropriate coordinates of the original summary points because the summary points are defined in terms of the ordered data values—first by using the ordered *x*-values to divide the data into thirds and then by using the ordered *x*-values and ordering the *y*-values to find medians within each third. We already have seen (in Section 2.4) that re-expressions on the ladder of powers preserve order. Thus, the coordinates of the summary points of the re-expressed data are simply the re-expressed coordinates of the original summary points.

The half-slope ratio is computed from the summary points. Thus we need not re-express all of the data; we can re-express the summary points alone and compute a new half-slope ratio. We can then explore a variety of re-expressions quickly and easily without having to re-express every data value for each try. However, (as Section 2.4 warned) when two data values have been averaged to compute a median for a summary point coordinate, we may prefer to re-express each of them and then average so we can be more accurate.

Example: Automobile Gasoline Mileage

Exhibit 5–11 reports mileage (in miles per gallon) and engine size (specifically, displacement in cubic inches) for thirty-two 1976-model automobiles. The data are plotted in Exhibit 5–12. The plot clearly bends in a direction that indicates a move down in the power of *x* or down for *y*. The half-slopes are −0.083 and −0.022, and their ratio is 0.268. We could try to re-express *x* or *y*, and the nature of the data suggests one re-expression. Gasoline mileage was actually estimated by driving a measured course and observing the amount of gasoline consumed—that is, by finding gallons used per mile. If we take the

Exhibit 5–11 Gas Mileage and Displacement for Some 1976-Model Automobiles

Automobile	mpg	Displacement
Mazda RX–4	21.0	160.0
Mazda RX–4 Wagon	21.0	160.0
Datsun 710	22.8	108.0
Hornet 4-Drive	21.4	258.0
Hornet Sportabout	18.7	360.0
Valiant	18.1	225.0
Plymouth Duster	14.3	360.0
Mercedes 240D	24.4	146.7
Mercedes 230	22.8	140.8
Mercedes 280	19.2	167.6
Mercedes 280C	17.8	167.6
Mercedes 450SE	16.4	275.8
Mercedes 450SL	17.3	275.8
Mercedes 450SLC	15.2	275.8
Cadillac Fleetwood	10.4	472.0
Lincoln Continental	10.4	460.0
Chrysler Imperial	14.7	440.0
Fiat 128	32.4	78.7
Honda Civic	30.4	75.7
Toyota Corolla	33.9	71.1
Toyota Corona	21.5	120.1
Dodge Challenger	15.5	318.0
AMC Javelin	15.2	304.0
Chevrolet Camaro Z–28	13.3	350.0
Pontiac Firebird	19.2	400.0
Fiat X1–9	27.3	79.0
Porsche 914–2	26.0	120.3
Lotus Europa	30.4	95.1
Ford Pantera L	15.8	351.0
Ferrari Dino 1973	19.7	145.0
Maserati Bora	15.0	301.0
Volvo 142E	21.4	121.0

Source: From data set supplied by Ronald R. Hocking. Used with permission.

Exhibit 5–12   Gas Mileage versus Displacement for Some 1976-Model Automobiles

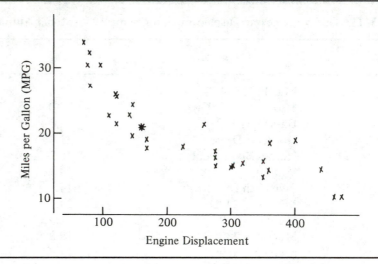

reciprocal of the miles per gallon data (the −1 power), which is down the ladder of powers, as we want, we obtain data in gallons per mile. This plot is straighter—the half-slope ratio is 0.46—but not entirely straight (see Exhibit 5–13).

The shape of the plot of gallons per mile against displacement shown in

Exhibit 5–13   Gallons per Mile versus Displacement for Some 1976-Model Automobiles

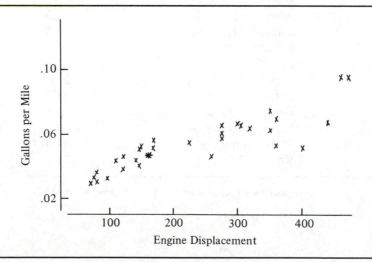

Exhibit 5–14 Gallons per Mile versus (Displacement)$^{-1/3}$

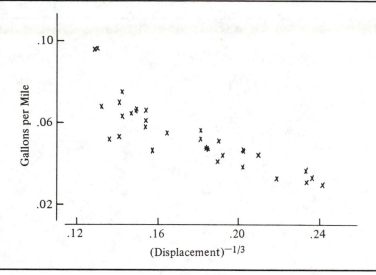

Exhibit 5–13 indicates a move down in *x*. We might try gallons per mile and
$\sqrt{}$(displacement), which is the ½ power. This pair of re-expressions yields a
half-slope ratio of 0.61—a value closer to 1.0 but still not satisfactory. If we
move to log(displacement), which is the zero power, the half-slope ratio is
0.81. One more step to 1/(displacement), which is the −1 power, seems to go
too far: The half-slope ratio is 1.43. Thus we know that some power between
−1 and 0 (the log) should do a good job. After a few more trials, we find that
the reciprocal cube root, the −⅓ power, does quite well. The half-slope ratio
for (mpg)$^{-1}$ versus (displacement)$^{-1/3}$ is 0.98—a value very close to the ideal
of 1.0. Displacement is measured in cubic inches; so the reciprocal cube root

Exhibit 5–15 Resistant Line for the Re-expressed Data of Exhibit 5–14

```
HALF-SLOPE RATIO = 1.0191
SLOPE 1: -.4063
SLOPE 2: -.3520
SLOPE 3: -.3752
SLOPE 4: -.3636
SLOPE 5: -.3751
FITTED LINE:
Y = .12 + -.375 X
```

Exhibit 5–16 Residuals versus (Displacement)$^{-1/3}$ for Line Fitted to Gallons per Mile in Exhibit 5–15

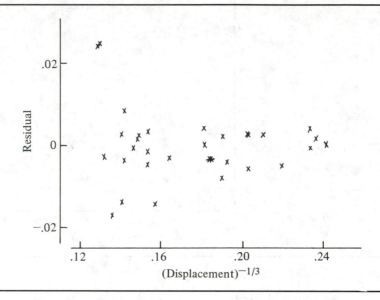

has simple units: 1/inches. Exhibits 5–14, 5–15, and 5–16 show the plot, the resistant line, and the residuals, respectively.

This example illustrates an important aspect of re-expressing data. Often, especially if both x and y are re-expressed, more than one pair of re-expressions will make a plot reasonably straight. In these situations we should use any available knowledge about the data to make a final choice. In this example we considered how mileage is measured and the units of displacement. Considerations of this nature keep us from automating re-expression entirely, although if our only goal were a straight plot, that could be done.

5.9   Interpreting Fits to Re-expressed *x-y* Data

While some re-expressions are easy to understand ("gallons per mile" is as natural as "miles per gallon"), often we have to take extra care in describing a line fit to re-expressed x or y data values. We noted at the beginning of this chapter that the intercept has the same units as the y-variable, and the slope is in "units of y per unit of x." If either x or y is re-expressed, we need to use the re-expressed units in interpreting the slope and intercept. Thus in the gas mileage example (Exhibit 5–15), the intercept could be interpreted as .12 gallons per mile, and the slope as -0.375 *gallons per mile* per *reciprocal inch of engine size*. Because we have re-expressed both x and y, the units of the slope are further away from the units of the original data. We can, however, check that the sign of the slope is reasonable—a smaller engine would have a larger reciprocal size and hence would use fewer gallons of gasoline per mile—and it is still easy to use a new engine size in predicting gasoline consumption.

When we have re-expressed y, an alternative interpretation can be found by inverting the re-expression to obtain a fit for y in its original units. Instead of the fitted linear equation $\sqrt{y} = a + bx$, we could consider the equivalent form

$$y = (a + bx)^2 = a^2 + 2abx + b^2x^2.$$

Instead of the fitted linear equation $\log(y) = a + bx$, we could consider the form $y = 10^{(a+bx)}$. Generally, whatever we gain by simplifying the expression of y, we lose by making the fitted equation more complex. We have, in the resistant line, a convenient technique for fitting a line to an x-y relationship. Re-expressions extend the power of this technique to cover a far wider range of x-y relationships without the need for new fitting methods.

The residuals from a line fit to re-expressed y-values must be computed in the re-expressed units. Thus the residuals in the gas mileage example are found from

$$\frac{1}{\text{observed mpg}} - [0.12 - 0.375(\text{disp})^{-1/3}]$$

and are in the re-expressed units, gallons per mile.

Sometimes, the first hint of a need to re-express y will be that the residuals would look better after re-expression. For example, often larger y-values are measured less precisely than smaller values. The residuals will then show a wedge pattern when plotted against x—that is, they will be more spread out at the x-values corresponding to large y-values, less spread out where the y-values (and the measurement fluctuations) were smaller. Re-expressing y by moving down the ladder of powers will often make the measurement fluctuations more comparable and make the residuals more evenly spread out. When a single re-expression of y both straightens the x-y relationship and evens up the residual pattern, we might have additional faith that it is a worthwhile re-expression, and we would rather use it to straighten the relationship than re-express the x-variable.

* 5.10   Resistant Lines and Least-Squares Regression

The resistant line is one of many ways to fit a linear model to *y*-versus-*x* data. The most common method is least-squares regression. Of course, these two methods will generally not yield the same slope and intercept estimates, but often the two sets of estimates will agree quite closely.

When our data contain outliers, or even when the distribution of the residuals—from either fitted line—has long tails, the resistant line is likely to differ more markedly from the regression line. The primary reason for this difference is that least-squares regression is not resistant to the effects of outliers.

When the distribution of the residuals is close to Gaussian and the data satisfy some other restrictions, least-squares regression permits us to make statistical inferences about the line. The resistant line is not yet accompanied by an inference procedure. However, if the data do not meet the conditions for regression, it is dangerous to draw inferences from a least-squares line. In such instances, the resistant-line technique is likely to provide a better description of the data.

Most statistical computer packages include programs for least-squares regression. When we are analyzing data with such a package, it is usually worthwhile to fit both a resistant line and a least-squares regression and compare the two lines. If they are similar, the regression line might be preferred for the inference calculations it allows. If the lines differ, the residuals from the resistant line may reveal the reason.

When we work by hand, we will usually prefer the resistant line because of its simpler calculations. When we use a computer, it is often helpful to fit a resistant line first. This allows us to (1) check that the *y*-versus-*x* relationship is linear, (2) find a re-expression to straighten the relationship if necessary, and (3) check the residuals for outliers. Once we are reassured that the data are well-behaved in these ways, we can fit a least-squares regression line.

5.11   Resistant Lines from the Computer

As we have seen, the computer can save us much calculating work in finding a resistant line and can print the slope of the fitted line at each step of the iteration. We must tell the programs which variables to treat as *x* and *y*. In addition, we should specify where the residuals are to be put. These specifications may not be necessary in some implementations. They are automatic in the BASIC programs.

The programs offer two modes of operation: verbose and silent. In verbose mode—the recommended mode for exploring data—each iteration of line polishing is reported. In silent mode, only the final fit is reported. As an additional option, the programs can be told to limit the number of polish iterations. The default limit is 10 iterations—usually more than enough. Some peculiar *x-y* data (especially when ties among *x*-values drastically reduce the size of the middle third) may require more iterations.

In addition to the resistant line and residuals, the programs also report the half-slope ratio for assessing the straightness of the *x-y* relationship. However, the program will attempt to fit a line even if the half-slope ratio indicates nonlinearity. It is up to the data analyst to recognize and treat this difficulty.

† 5.12 Algorithms

The programs begin by dividing the batch into thirds and finding summary points. The algorithm to do this ensures that points with the same *x*-values will be assigned to the same region and that no region will have too few points. (If one of the outer regions has fewer than 3 points, the line will not be resistant.) If only two distinct regions can be defined, the programs proceed with them. If even this is impossible, the programs report the error.

Resistant-line polishing iterates until the slope estimate is correct to at least four digits. (The algorithm does this by keeping an upper and a lower bound on the correct slope.) The user must supply a maximum for the number of steps, in case the process fails to converge. If this happens, the programs return the last bounds on the slope. Otherwise, they return the final slope estimate and an intercept estimate chosen to make the median of the residuals zero.

FORTRAN

The FORTRAN program for resistant line is a single subroutine, RLINE. When in verbose mode (TRACE set .TRUE.), it writes a report of each iteration. However, in both verbose and silent modes, the program returns the final fit without printing it. Thus the calling program is responsible for printing the results. This makes it possible to use the resistant-line subroutine as a part of a larger program. To request a resistant line for data values (x, y) in the parallel arrays X( ) and Y( ), use the statement

CALL RLINE(X, Y, N, RESID, WORK, NSTEPS, SLOPE, LEVEL, LLS, LUS, TRACE,
    LHSLOP, RHSLOP, HSRTIO, ERR)

The arguments are as follows:

X( ),Y( )	are N-long arrays holding the data pairs;
N	is the number of data values;
RESID( )	is an N-long array in which residuals are returned;
WORK( )	is an N-long scratch array;
NSTEPS	is the maximum number of polish iterations permitted;
SLOPE, LEVEL	are REAL-valued variables, which return b and a;
LLS, LUS	are the "last lower slope" and "last upper slope"—return zero if the iteration has converged, otherwise return the last bounds on the slope;
TRACE	is a LOGICAL variable, set .TRUE. to report each iteration or .FALSE. to just pass back the solution;
LHSLOP, RHSLOP, HSRTIO	are the left half-slope, right half-slope, and their ratio, RHSLOP/LHSLOP (returned by the subroutine to aid in assessing straightness);
ERR	is the error flag, whose values are

	0	normal
	51	$N < 6$—too few data values
	52	NSTEPS = 0—no iteration requested
	53	all x-values equal—no line possible
	54	split is too uneven for resistance.

BASIC

The BASIC program for resistant-line fitting expects N (x, y) pairs in the parallel arrays X( ) and Y( ). It returns coefficients in B0 and B1, and residuals in

R( ). Before the first fitting step, the program prints the half-slope ratio. For version V1 = 1 the program requests a maximum iteration limit and reports only the final fit; otherwise it reports the slope at each iteration. In this verbose mode, the output format is modified to round the slope so that the last two digits of the number printed are the only ones likely to have changed since the previous iteration. This makes it easy to judge the precision of the slope estimate as the iteration proceeds.

The program returns X( ) and Y( ) sorted on X( ) and returns the residuals (also sorted on X( )) in R( ). The program uses the defined functions, the pair sorting subroutine, and the sorting subroutines.

Reference

Lea, A.J. 1965. "New Observations on Distribution of Neoplasms of Female Breast in Certain European Countries." *British Medical Journal* 1:488–490.

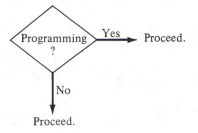

```
5000 REM COMPUTE AND PRINT RESISTANT LINE FOR N PAIRS (X,Y)
5010 REM   IN X(), Y(). ON EXIT, X() AND Y() HOLD ORIGINAL DATA
5020 REM   SORTED ON X(); R() HOLDS RESIDUALS SORTED ON X().
5030 REM    IF V1>1 PRINTS APPROXIMATIONS AT EVERY STEP.
5040 REM DEFAULT MAX#ITERATIONS=10, TOL = 1.0E-4

5050 LET J9 = 10
5060 LET T0 = 1.0E - 4 * 0.5
5070 IF N > 5 THEN 5100
5080 PRINT "N<=5"
5090 RETURN
5100 IF V1 > 0 THEN 5140
5110 PRINT TAB(M0);"MAXIMUM # ITERATIONS";
5120 INPUT J9
5130 LET V1 = ABS(V1)

5140 REM   SORT ON X CARRYING Y

5150 GOSUB 1200

5160 REM   **FIND EDGES OF THE THIRDS**

5170 LET E1 = (N + 1) / 2
5180 LET E3 = E1
5190 LET M = FNN(E1)
5200 FOR E1 = INT(E1) TO 1 STEP - 1
5210   IF X(E1) < M THEN 5250
5220 NEXT E1

5230 REM   ALL VALUES ARE TIED FROM MEDIAN TO LOW END

5240 LET E1 = 0
5250 FOR E3 = INT(E3 + .5) TO N
5260   IF X(E3) > M THEN 5350
5270 NEXT E3

5280 REM   ALL VALUES ARE TIED FROM MEDIAN TO HIGH END

5290 IF E1 > 0 THEN 5320
5300 PRINT TAB(M0);"X IS CONSTANT--NO FIT POSSIBLE"
5310 RETURN

5320 REM   ONLY 2 GROUPS

5330 LET E3 = E1 + 1
5340 GO TO 5380
5350 IF E1 > 0 THEN 5380
5360 LET E1 = E3 - 1

5370 REM   NOW PLACE THE THIRDS

5380 IF E1 <= 3 THEN 5470
5390 LET T1 = INT((N + 1) / 3)
5400 LET X1 = X(T1)
```

```
5410 REM   IF T1 > E1 THEN LOOP IS SKIPPED AND E1 = E9

5420 LET E9 = E1
5430 FOR E1 = T1 TO E9
5440   IF X(E1 + 1) <> X1 THEN 5470
5450 NEXT E1
5460 LET E1 = E9

5470 REM   PLACE HIGH THIRD

5480 IF E3 >= N - 2 THEN 5570
5490 LET T3 = N - T1 + 1
5500 LET X3 = X(T3)
5510 LET E9 = E3

5520 REM   IF T3 < E3 THEN LOOP IS SKIPPED AND E3 = E9

5530 FOR E3 = T3 TO E9 STEP - 1
5540   IF X(E3 - 1) <> X3 THEN 5570
5550 NEXT E3
5560 LET E3 = E9

5570 REM **NOW E1 AND E3 ARE INNER EDGES OF OUTER THIRDS**
5580 REM **SET UP FOR FITTING**

5590 LET N1 = E1
5600 LET N3 = N - E3 + 1
5610 LET N2 = N - N1 - N3
5620 LET N9 = N
5630 IF N2 < 2 THEN 5720
5640 IF N1 > 2 THEN 5700
5650 IF N3 > 2 THEN 5680
5660 PRINT TAB(M0);"NOT ENOUGH DIFFERENT X-VALUES"
5670 RETURN
5680 LET E1 = E3 - 1
5690 GO TO 5720
5700 IF N3 > 2 THEN 5780
5710 LET E3 = E1 + 1

5720 REM   ONLY 2 GROUPS

5730 LET N1 = E1
5740 LET N2 = 0
5750 LET N3 = N - E3 + 1
5760 LET X2 = 0
5770 LET Y2 = 0

5780 REM   CONTINUE

5790 LET M1 = (N1 + 1) / 2
5800 LET M2 = (N2 + 1) / 2
5810 LET M3 = (N3 + 1) / 2
```

```
5820 REM   GET X-MEDIANS (STILL SORTED ON X)

5830 LET X1 = FNN(M1)
5840 LET X4 = X(E1)
5850 IF N2 = 0 THEN 5870
5860 LET X2 = FNN(E1 + M2)
5870 LET X3 = FNN(E3 + M3 - 1)
5880 LET X5 = X(E3)
5890 LET D8 = X3 - X1

5900 REM   GET Y-MEDIANS

5910 LET N = N1
5920 GOSUB 3300
5930 LET Y1 = FNM(M1)
5940 LET Y4 = W(1)
5950 LET Y5 = W(N1)
5960 IF N2 = 0 THEN 6010
5970 LET J1 = E1 + 1
5980 LET J2 = E1 + N2
5990 GOSUB 3340
6000 LET Y2 = FNM(M2)
6010 LET J1 = E3
6020 LET J2 = N9
6030 GOSUB 3340
6040 LET Y3 = FNM(M3)
6050 LET Y6 = W(1)
6060 LET Y7 = W(N)

6070 REM   ON FIRST ITERATION, REPORT ON BEND

6080 IF V1 < 2 THEN 6140
6090 IF N2 = 0 THEN 6140
6100 LET B6 = (Y3 - Y2) / (X3 - X2)
6110 LET B5 = (Y2 - Y1) / (X2 - X1)
6120 IF ABS(B5) <= E0 THEN 6140
6130 PRINT TAB(M0);"HALF-SLOPE RATIO = ";B6 / B5

6140 REM   FIRST 2 STEPS OF POLISH TO START

6150 LET B2 = (Y3 - Y1) / D8
6160 GOSUB 7040
6170 LET B1 = B2
6180 LET D1 = D2
6190 LET R0 = 4
6200 IF V1 < 2 GO TO 6220
6210 PRINT TAB(M0);"SLOPE 1: "; FNR(B1)
6220 LET B3 = B2
6230 LET D6 = D2 / D8
6240 IF ABS(D6) < E0 THEN 6850
6250 LET B2 = B3 + D6
6260 GOSUB 7040
6270 IF SGN(D2) <> SGN(D1) THEN 6320
6280 LET D6 = D6 + D6
6290 LET B1 = B2
```

```
6300 LET D1 = D2
6310 GO TO 6250
6320 IF V1 < 2 THEN 6340
6330 PRINT TAB(M0);"SLOPE 2: "; FNR(B2)

6340 REM ITERATION BASED UPON ZEROIN (SEE FORSYTH, MALCOM, & MOLER)

6350 LET J8 = 2
6360 LET B3 = B1
6370 LET D3 = D1
6380 LET B4 = B2 - B1
6390 LET B5 = B4
6400 IF ABS(D3) >= ABS(D2) GO TO 6470
6410 LET B1 = B2
6420 LET B2 = B3
6430 LET B3 = B1
6440 LET D1 = D2
6450 LET D2 = D3
6460 LET D3 = D1
6470 IF J8 > J9 GO TO 6820

6480 REM T1,T2,T3 USED FOR TOLERANCES FROM HERE ON

6490 LET T1 = 2 * E0 * ABS(B2) + T0
6500 LET B6 = 0.5 * (B3 - B2)
6510 IF ABS(B6) <= T1 GO TO 6850
6520 IF D2 = 0 GO TO 6850

6530 REM   TRY AGAIN

6540 LET D4 = D2 / D1
6550 LET T2 = 2 * B6 * D4
6560 LET T3 = 1 - D4
6570 IF T2 < 0 GO TO 6590
6580 LET T3 = - T3
6590 LET T2 = ABS(T2)
6600 IF 2 * T2 >= 3 * B6 * T3 - ABS(T1 * T3) GO TO 6660
6610 IF T2 >= ABS(0.5 * B5 * T3) GO TO 6660
6620 LET B5 = B4
6630 LET B4 = T2 / T3
6640 GO TO 6680

6650 REM   BISECT FOR NEXT TRY

6660 LET B4 = B6
6670 LET B5 = B4

6680 REM   SECANT RULE

6690 LET B1 = B2
6700 LET D1 = D2
6710 LET B2 = B2 + B4
6720 IF ABS(B4) > T1 GO TO 6740
6730 LET B2 = B1 + T1 * SGN(B6)
6740 LET J8 = J8 + 1
```

```
6750 REM   REPORT STEP

6760 IF V1 < 2 GO TO 6790
6770 LET R0 = - FNF( FNL(B6)) + 1
6780 PRINT TAB(M0);"SLOPE ";J8;": "; FNR(B2)
6790 GOSUB 7070
6800 IF SGN(D2) = SGN(D3) GO TO 6360
6810 GO TO 6400
6820 PRINT "FAILED TO CONVERGE AFTER ";J9;" ITERATIONS."
6830 PRINT TAB(M0);B2;" <= B <= ";B3

6840 REM   COMPUTE INTERCEPT AND RESIDUALS ANYWAY
6850 REM   EXIT -- PRINT FINAL EQUATION

6860 LET N = N9
6870 FOR I = 1 TO N
6880   LET W(I) = Y(I) - B2 * X(I)
6890   LET R(I) = W(I)
6900 NEXT I
6910 GOSUB 1000
6920 LET B0 = FNM((N + 1) / 2)
6930 PRINT
6940 PRINT TAB(M0);"FITTED LINE:"
6950 PRINT "Y =";
6960 PRINT FNR(B0);
6970 IF ABS(D2) > E0 THEN 6990
6980 LET R0 = 7
6990 PRINT " + "; FNR(B2);" X"
7000 FOR I = 1 TO N
7010   LET R(I) = R(I) - B0
7020 NEXT I
7030 RETURN

7040 REM   SUBROUTINE TO FIND MEDIAN RESIDUALS AND THEIR DIFFERENCE.
7050 REM ENTERED WITH TRIAL SLOPE IN B2
7060 REM PUTS DIFFERENCE BETWEEN LEFT AND RIGHT MEDIAN RESIDS IN D2

7070 LET N = N1
7080 FOR I = 1 TO N1
7090   LET W(I) = Y(I) - X(I) * B2
7100 NEXT I
7110 GOSUB 1000
7120 LET Z1 = FNM(M1)
7130 LET N = 0
7140 FOR I = E3 TO N9
7150   LET N = N + 1
7160   LET W(N) = Y(I) - X(I) * B2
7170 NEXT I
7180 GOSUB 1000
7190 LET D2 = FNM(M3) - Z1
7200 RETURN
```

```
      SUBROUTINE RLINE(X, Y, N, RESID, WORK, NSTEPS, SLOPE, LEVEL, LLS,
     1 LUS, TRACE, LHSLOP, RHSLOP, HSRTIO, ERR)
C
      INTEGER N, NSTEPS, ERR
      REAL X(N), Y(N), RESID(N), WORK(N), SLOPE, LEVEL, LLS, LUS
      REAL LHSLOP, PHSLOP, HSRTIO
      LOGICAL TRACE
C
C FOR THE DATA (X(1), Y(1)), ... , (X(N), Y(N)), FIT THE STRAIGHT LINE
C    Y = LEVEL + SLOPE * X  + RESID
C BY THE "RESISTANT LINE" TECHNIQUE.
C ITERATES FOR  NSTEPS  STEPS OF UNTIL THE SLOPE IS CORRECT TO 4
C DIGITS.  1/TOL SPECIFIES THE NUMBER OF DIGITS REQUIRED.
C IF CONVERGENCE NCT ATTAINED AFTER NSTEPS STEPS, LLU AND LLS WILL
C RETURN THE LAST LOWER AND UPPER BOUNDS ON THE CORRECT SLOPE,
C OTHERWISE THEY WILL RETURN ZERO.
C THIS METHOD WILL NOT WORK FOR N .LE. 5, AND IT WILL NOT BE FULLY
C RESISTANT FOR N .LE. 7.  IF SEVERAL X-VALUES ARE TIED, N SHOULD BE
C STILL LARGER TO GUARANTEE RESISTANCE.
C THE PROGRAM ALSO COMPUTES THE APPROXIMATE SLOPE OF THE LEFT HALF
C AND OF THE RIGHT HALF OF THE DATA IN  LHSLOP  AND  RHSLOP.
C THEIR RATIO, RETURNED IN  HSRTIO, IS A MEASURE OF THE STRAIGHTNESS
C OF THE X-Y RELATIONSHIP.
C IF  TRACE  IS .TRUE. ON ENTRY, THE HALFSLOPE RATIO WILL BE PRINTED
C AND A REPORT WILL BE PRINTED AFTER EACH STEP OF THE ITERATION.
C
C COMMCN
C
      COMMON /NUMBRS/ EPSI, MAXINT
      REAL EPSI, MAXINT
      COMMON /CHRBUF/P, PMAX, PMIN, OUTPTR, MAXPTR, OUNIT
      INTEGER P(130), PMAX, PMIN, OUTPTR, MAXPTR, OUNIT
C
C LOCAL VARIABLES
C
      INTEGER I, MPT1, MPT2, MPT3, N1, N2, N3, L3RD, R3RD
      INTEGER MXLO3, MNHI3, FROM, TO, STEPNO, MPTX, MPTY
      REAL X1, X2, X3, Y1, Y2, Y3, XMED, DSLOPE, TOL, TOL1
      REAL SLOPE1, SLOPE2, SLOPE3, DELTX, DR1, DR2, DR3
      REAL OLDDS, DDR, NUMTOR, DENOM, DSD2
C
C FUNCTIONS
C
      REAL RL3MED, DELTR, MEDIAN
C
C 1/TOL  SPECIFIES NUMBER OF RELIABLE SLOPE DIGITS REQUIRED
C
      TOL = 1.0E-4
      LLS = 0.0
      LUS = 0.0
C
      IF(N .GT. 5) GOTO 5
      ERR = 51
      GOTO 999
    5 IF (NSTEPS .GT. 0) GOTO 10
      ERR = 52
      GOTO 999
C
C DIVIDE INTO THIRDS ON X
C   FIRST CHECK FOR TIES
```

```
C
      10 CALL PSORT(X, Y, N, ERR)
         IF (ERR .NE. 0) GOTO 999
C
         MPT2 = (N/2) +1
         MPT1 = N - MPT2 + 1
         XMED = (X(MPT1) + X(MPT2))/2.0
C
C LOOK FOR FIRST VALUE NOT TIED WITH MEDIAN.  IT IS THE MAX POSSIBLE
C  LOW THIRD CUT.
C
         MXLO3 = MPT2
      20 MXLO3 = MXLO3 -1
         IF(X(MXLO3) .NE. XMED) GOTO 30
         IF( MXLO3 .GT. 1) GOTO 20
C
C FALL THROUGH HERE IF ALL TIED FROM LOW END TO MEDIAN
C
         MXLO3 = 0
C
C LOOK FOR MINIMUM POSSIBLE HIGH THIRD CUT
C
      30 MNHI3 = MPT1
      40 MNHI3 = MNHI3 + 1
         IF( X(MNHI3) .NE. XMED ) GOTO 60
         IF ( MNHI3 .LT. N ) GOTO 40
C
C FALL THROUGH HERE IF ALL TIED FROM MEDIAN TO HIGH END.
C
         MNHI3 = N+1
         IF (MXLO3 .NE. 0 ) GOTO 50
C
C ALL TIED HIGH TO LOW -- CANT FIND A SLOPE
C
         ERR = 53
         GOTO 999
C
C ONLY TWO "THIRDS"
C
      50 MNHI3 = MXLO3 + 1
         GOTO 70
      60 IF( MXLO3 .NE. 0 ) GOTO 70
C
C LOW THIRD EMPTY
C
         MXLO3 = MNHI3 - 1
      70 CONTINUE
C
C NOW PLACE THE THIRDS
C  GET FAVORED LOW SPLIT POINT
C
         MPT1 = (N+1)/3
         X1 = X(MPT1)
C
C DONT SPLIT TIES.  FAVOR LARGER OUTER THIRDS.
C
         L3RD = MXLO3
         IF( MPT1 .GT. MXLO3) GOTO 90
         L3RD = MPT1
```

```
    80 L3RD = L3RD + 1
       IF( X(L3RD) .EQ. X1 ) GOTO 80
       L3RD = L3RD - 1
C
C  NOW THE HIGH THIRD
C
    90 MPT3 = N - MPT1 + 1
       X3 = X(MPT3)
C
C  DONT SPLIT TIES.  FAVOR LARGER OUTER THIRDS.
C
       R3RD = MNHI3
       IF( MPT3 .LE. MNHI3) GOTO 110
       R3RD = MPT3
   100 R3RD = R3RD - 1
       IF (X(R3RD) .EQ. X3 ) GOTO 100
       R3RD = R3RD + 1
   110 CONTINUE
C
C  NOW L3RD AND R3RD POINT TO INNER EDGES OF OUTER THIRDS.
C
C  CHECK IF THIRDS ARE BIG ENOUGH FOR RESISTANCE.
C
       N1 = L3RD
       N3 = N - R3RD + 1
       N2 = N - N1 - N3
       IF ((N1 .GT. 2) .OR. (N3 .GT. 2)) GOTO 120
C
C  IF N = 7 AND SPLIT IS 2 - 3 - 2, STICK WITH IT.
C
       IF ((N1 .EQ. 2) .AND. (N2 .EQ. 3) .AND. (N3 .EQ. 2)) GOTO 140
       ERR = 54
       GOTO 999
   120 IF ((N1 .GT. 2) .AND. (N3 .GT. 2)) GOTO 140
C
C  ONLY 2 THIRDS ARE BIG ENOUGH -- REGROUP AND WORK WITH 2.
C
       IF (N1 .LE. 2) L3RD = R3RD - 1
       IF (N3 .LE. 2) R3RD = L3RD + 1
   130 N1 = L3RD
       N2 = 0
       N3 = N - R3RD + 1
       X2 = 0.0
       Y2 = 0.0
C
   140 CONTINUE
C
C  SET UP FOR FITTING
C
C  GET X MEDIANS
C
       MPT1 = (N1+1)/2
       MPT2 = (N2+1)/2
       MPT3 = (N3+1)/2
       MPTY = N1 - MPT1 + 1
       X1 = (X(MPT1) + X(MPTY))/2.0
       MPTX = N1 + MPT2
```

```
          MPTY = N1 + N2 - MPT2 + 1
          IF(N2 .NE. 0) X2 = (X(MPTX) + X(MPTY))/2.0
          MPTX = N1 + N2 + MPT3
          MPTY = N - MPT3 + 1
          X3 = (X(MPTX) + X(MPTY))/2.0
          DELTX = X3-X1
          IF (ABS(DELTX) .LT. EPSI) DELTX = SIGN(EPSI, DELTX)
C
C   Y - MEDIANS
C
          Y1 = RL3MED(Y, N, 1, L3RD, WORK, ERR)
          FROM = L3RD + 1
          TO = R3RD - 1
          IF(N2 .NE. 0) Y2 = RL3MED(Y, N, FROM, TO, WORK, ERR)
          Y3 = RL3MED(Y, N, R3RD, N, WORK, ERR)
          IF (ERR .NE. 0) GOTO 999
C
C   COMPUTE HALF-SLOPE RATIO TO CHECK STRAIGHTNESS OF Y ON X.
C   REPORT IF TRACE IS .TRUE. ELSE JUST RETURN RESULTS.
C
          IF( N2 .EQ. 0 ) GO TO 170
          LHSLOP = (Y2 - Y1)/(X2 - X1)
          RHSLOP = (Y3 - Y2)/(X3 - X2)
          IF (ABS(LHSLOP) .GT. EPSI) GO TO 160
          HSRTIO = 0.0
          GO TO 170
  160     HSRTIO = RHSLOP/LHSLOP
          IF(TRACE) WRITE(OUNIT, 5002) LHSLOP, RHSLOP, HSRTIO
 5002     FORMAT(1X, 19HSTRAIGHTNESS CHECK./1X, 18H LEFT HALF-SLOPE =,
     2    F12.6, 19H RIGHT HALF-SLOPE =, F12.6/10X, 8H RATIO =, F12.6//)
  170     CONTINUE
C
C   FIRST 2 SLOPES WITHOUT ITERATING
C
          STEPNO = 1
          SLOPE1 = (Y3 - Y1)/DELTX
          DR1 = DELTR(X, Y, N, RESID, L3RD, R3RD, SLOPE1, WORK, ERR)
          IF(ERR .NE. 0) GO TO 999
          DSLOPE = DR1/DELTX
          IF (TRACE) WRITE(OUNIT, 5000) STEPNO, SLOPE1
 5000 FORMAT(1X, 6HSLOPE ,I3,2H: ,F12.6)
          STEPNO = 2
          SLOPE2 = SLOPE1 + DSLOPE
          SLOPE3 = SLOPE1
  180 DR2 = DELTR(X, Y, N, RESID, L3RD, R3RD, SLOPE2, WORK, ERR)
          IF(ERR .NE. 0) GO TO 999
          IF(DR2 .EQ. 0.0) GO TO 290
C   FIND SECOND SLOPE WITH OPPOSITE-SIGN RESIDUAL DIFFERENCE
          IF(SIGN(1.0, DR2) .NE. SIGN(1.0, DR1)) GO TO 190
          SLOPE1 = SLOPE2
          DR1 = DR2
          SLOPE2 = SLOPE3 + DSLOPE
          DSLOPE = DSLOPE + DSLOPE
          GO TO 180
  190 IF (TRACE) WRITE(OUNIT, 5000) STEPNO, SLOPE2
          ADR = ABS(DR2)
C
C   ITERATION IS BASED UPON THE ALGORITHM  ZEROIN  (SEE FORSYTHE,
C   MALCOM, AND MOLER P161 FF.)
```

```
C
  220 SLOPE3 = SLOPE1
      DR3 = DR1
      DSLOPE = SLOPE2 - SLOPE1
      OLDDS = DSLOPE
  230 IF( ABS(DR3) .GE. ABS(DR2) ) GO TO 240
      SLOPE1 = SLOPE2
      SLOPE2 = SLOPE3
      SLOPE3 = SLOPE1
      DR1 = DR2
      DR2 = DR3
      DR3 = DR1
C
C  TEST CONVERGENCE
C
  240 IF( STEPNO .GE. NSTEPS ) GO TO 285
      TOL1 = 2.0 * EPSI * ABS(SLOPE2) + 0.5 * TOL
      DSD2 = .5 * (SLOPE3 - SLOPE2)
      IF(ABS(DSD2) .LE. TOL1) GO TO 290
      IF(DR2 .EQ. 0.0) GO TO 290
C
C  TRY AGAIN
C
      DDR = DR2/DR1
      NUMTOR = 2.0 * DSD2 * DDR
      DENOM = 1.0 - DDR
      IF( NUMTOR .GT. 0.0 ) DENOM = -DENOM
      NUMTOR = ABS(NUMTOR)
      IF((2.0 * NUMTOR) .GE. (3.0 * DSD2 * DENOM - ABS(TOL1 * DENOM)))
     1    GO TO 270
      IF( NUMTOR .GE. ABS(0.5 * OLDDS * DENOM) ) GO TO 270
      OLDDS = DSLOPE
      DSLOPE = NUMTOR/DENOM
      GO TO 280
C
C BISECT
C
  270 DSLOPE = DSD2
      OLDDS = DSLOPE
C
  280 SLOPE1 = SLOPE2
      DR1 = DR2
      IF( ABS(DSLOPE) .GT. TOL1 ) SLOPE2 = SLOPE2 + DSLOPE
      IF( ABS(DSLOPE) .LE. TOL1 ) SLOPE2 = SLOPE2 + SIGN(TOL1,DSD2)
      STEPNO = STEPNO + 1
      IF(TRACE) WRITE(OUNIT, 5000) STEPNO, SLOPE2
      DR2 = DELTR(X, Y, N, RESID, L3RD, R3RD, SLOPE2, WORK,ERR)
      IF( ERR .NE. 0) GO TO 999
      IF( (DR2 * (DR3/ABS(DR3))) .GT. 0.0 ) GO TO 220
      GO TO 230
C
C  RAN OUT OF STEPS
C
  285 LLS = AMIN1(SLOPE1, SLOPE3)
      LUS = AMAX1(SLOPE1, SLOPE3)
      GO TO 999
C
C  EXIT
```

```
C
   290 SLOPE = SLOPE2
       DO 300 I = 1, N
          WORK(I) = Y(I) - SLOPE * X(I)
   300 CONTINUE
       CALL SORT( WORK, N, ERR)
       IF(ERR .NE. 0) GO TO 999
       LEVEL = MEDIAN(WORK, N)
       DO 310 I = 1, N
          RESID(I) = Y(I) - SLOPE*X(I) - LEVEL
   310 CONTINUE
   999 RETURN
       END
       REAL FUNCTION RL3MED(Y, N, FROM, TO, WORK, ERR)
C
C   RETURNS THE MEDIAN OF THE NUMBERS FROM Y(FROM) TO Y(TO), INCLUSIVE.
C
       INTEGER N, FROM, TO, ERR
       REAL Y(N), WORK(N)
C
C   LOCAL VARIABLES
C
       INTEGER I, J
C
C   FUNCTION
C
       REAL MEDIAN
C
       J = 0
       DO 10 I = FROM, TO
          J = J+1
          WORK(J) = Y(I)
    10 CONTINUE
       CALL SORT(WORK, J, ERR)
       IF (ERR .NE. 0 ) GOTO 999
       RL3MED = MEDIAN(WORK, J)
   999 RETURN
       END
       REAL FUNCTION DELTR(X, Y, N, RESID, L3RD, R3RD, SLOPE, WORK, ERR)
C
C   RETURNS THE DIFFERENCE BETWEEN THE MEDIAN RESIDUALS IN THE LEFT AND
C   RIGHT 3RDS OF THE DATA FOR A LINE WITH SPECIFIED SLOPE.
C
       INTEGER N, L3RD, R3RD, ERR
       REAL X(N), Y(N), RESID(N), WORK(N), SLOPE
C
       INTEGER I
C
C   FUNCTION
C
       REAL RL3MED
C
       DO 10 I = 1, N
          RESID(I) = Y(I) - SLOPE * X(I)
    10 CONTINUE
       DELTR = RL3MED(RESID, N, R3RD, N, WORK, ERR)
      2             - RL3MED(RESID, N, 1, L3RD, WORK, ERR)
       RETURN
       END
```

Chapter 6

Smoothing Data

The two previous chapters have presented techniques for plotting y-versus-x data and for summarizing such data with a resistant line. Often it is useful to search for patterns much more general than a straight line. When the x-values are equally spaced or almost equally spaced, we might ask only that y change smoothly from point to point along the x-axis. This chapter presents techniques for discovering and summarizing smooth data patterns.

6.1   Data Sequences and Smooth Summaries

When the x-values are equally spaced, their structure is so simple and regular that y often receives most of the attention. Lists of such data may even omit the x-values in favor of reporting the interval at which the data were recorded. *data sequence* We refer to such y-values as a ***data sequence***. Examples are the monthly rate of unemployment, the daily high and low temperatures at a weather station, and the number of votes cast in each U.S. presidential election.

When the sequence comes about by recording a value for each

time series   successive time interval, as in these examples, the *y*-values are known as a ***time series***. (Sometimes this term is reserved for such data sequences in which many consecutive values are available.) However, the order of data values in a sequence need not be defined by time. We might consider the sequence of birthrates as mother's age increases, heart-attack frequencies ordered by patient's weight, or the differences between low and high tide heights at points along a shoreline ordered by latitude. Data sequences are thus a specialized form of (x, y) data in which the values of *x* are important primarily for the order they specify—in time, in space, or whatever. Nevertheless, the terminology of time series is well suited to atemporal sequences as well. We might, for example, refer to a data value "earlier than" or "previous to" another value even if the ordering were not temporal. We therefore denote the order-defining value by *t* rather than *x* and often write it as a subscript to the variable *y*. Any data sequence can thus be represented as a sequence of values, y_t, ordered by *t*.

While the techniques in this chapter are usually applied to data whose *t*-values are evenly spaced, the essential feature of data sequences is that their *t*-values are in order. Sometimes we can take a fairly lax attitude toward the details of the spacing, provided that the spacing is not too irregular. Thus, as long as *t* defines an order, we may be able to use these techniques.

The Smooth and the Rough

In Chapter 5 we found it useful to treat a resistant line as a simple description of a *y*-versus-*x* relationship and to separate the data values into

$$\text{data} = \text{fit} + \text{residual}.$$

Such a separation can be useful even when the fit is not described by a formula. All we require is that the fit be a simple, well-structured description of the data and, ideally, that it capture much of the underlying pattern of the data.

Usually our attempts at a simple fit are smooth curves. When working by hand, we might plot the sequence of *y*-values against their corresponding *x*-values and sketch in a freehand curve. With such a curve we would try to capture the large-scale behavior of the data sequence—that is, where the sequence rises, where it falls, and whether it shows regularities or cycles (for example, greater sales in December of every year). Small-scale fluctuations, such as isolated data values out of line or small, rapidly changing oscillations, would then appear in the residuals.

However, if we want a simple fit to be reproducible or to be produced by computer, we must define the operations precisely. These smoothing operations usually summarize consecutive, overlapping segments of the sequence defined by *t*—for example, the first five data values, then the second through the sixth, and so on. Because the summarized segments overlap, the summaries change smoothly. The ***data smoothers*** discussed in this chapter use medians and averages to summarize the overlapping segments. The fit that these smoothers produce need not follow any specific formula; it is only required to be smooth. Therefore, we call it the ***smooth***. By contrast, we call the residuals the ***rough***. Thus we can write

*data
smoothers*

*smooth
rough*

$$data = smooth + rough.$$

The smooth and the rough, like the data values, are sequences ordered by *t*.

Note that (as in fitting lines) we may be more interested in the residuals, or rough, than in the fit, or smooth. One unfortunate consequence of the tradition that names these techniques "data smoothers" is that it may encourage some analysts to forget the importance of the rough.

Example: Daily Cow Temperatures

Exhibit 6–1 shows the body temperature of a cow measured at 6:30 A.M. on 75 consecutive days by a telemetric thermometer. This device is implanted in the cow and sends radio "chirps" to a nearby receiver. The higher the temperature, the faster the chirping. The data in Exhibit 6–1 are counts of chirps in a 5-minute interval on successive mornings. A dairy farmer might use a cow's temperature to help predict periods of fertility, which are usually associated with temperature peaks. It is difficult to see any pattern in Exhibit 6–1. We cannot tell whether the occasional high values are really at the peaks of temperature cycles or are just odd data values.

Exhibit 6–2 plots the smoothed sequence using one of the smoothers discussed in this chapter. The simplification is striking. In Exhibit 6–2, the *y*-values clearly rise and fall in 15- to 20-day cycles. Some of the higher values in Exhibit 6–1 do appear to be at peaks of cycles, but others just seem out of line. Cycles of about 15 to 20 days are consistent with the typical bovine reproductive cycle and may be related to changing hormone levels. Points out of line in the smooth sequence may indicate important events in the fertility of the cow or may simply have been recorded on a morning when the animal was either unusually active or sluggish. The steady slow decline in chirp frequency

Exhibit 6–1 Temperature of a Cow (in chirps per 5 minutes − 800) at 6:30 A.M. on 75
Consecutive Mornings. (Chirping rate transmitted is proportional to temperature.)

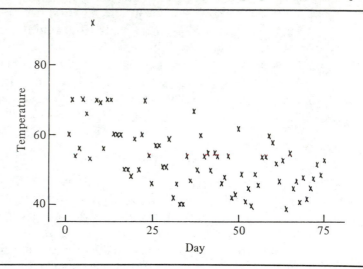

turns out to be due to the battery in the transmitter running down gradually.
The kind of display shown in Exhibit 6–2 is much more likely to be useful to
the farmer or veterinarian than is the display of the original data as in Exhibit
6–1. We will return to this example after learning more about how the
smoothing was done.

Exhibit 6–2 Cow Temperatures Smoothed

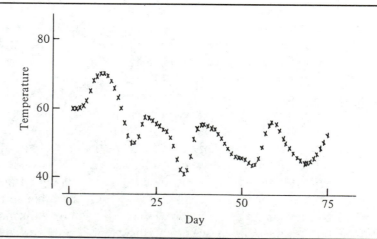

6.2   Elementary Smoothers

The fundamental property of a smooth sequence is that each data value is much like its neighbors; so changes do not take place suddenly. One simple way to achieve this is to replace each y-value with the median of three y-values—itself, its predecessor, and its successor. A y-value that is out of step with its neighbors will be replaced by one or the other of them, whichever is closer.

Running Medians

running-median smoothers

Because medians of three cannot correct for two outliers in a row, we may choose to take in more of the data. We can base each median in the smooth on five points instead of three by looking two points earlier and two points later than the y-value being modified. These two methods are examples of *running-median smoothers*, so named because we "run" along the data sequence and find the median of the three or five data values near each point.

For medians of three, the initial data value in the sequence poses a problem since it is not in the middle of three data values. For now, we just copy it without any modification. Of course, the same is true of the final data value, and we copy it for the smooth as well. For medians of five, the *two* data values at each end of the sequence are difficult to smooth. We copy the end values, but we use a median of three to smooth the second and next-to-last values.

After smoothing the rest of the sequence, we may want to modify the end values rather than just copy them. Section 6.4 discusses one useful method for smoothing the endpoints.

To show how running medians work, Exhibit 6–3 plots the first 30 days of the cow-temperature sequence, and Exhibits 6–4 and 6–5 show smooths of the data by running medians of three and five. While the smooth sequences are similar, they differ in recognizable ways: Generally the medians of five are more smooth but less like the original data sequence.

Each of these running-median smoothers can be computed easily by hand, but both are fairly heavy-handed in their effects on data sequences. Running medians of four consecutive data values are slightly gentler. Unlike smoothers that select the middle-sized data value of three or five, a running median of four values ignores the largest and smallest values in each segment of four and averages the two middle-sized values. Note that the values selected for averaging are of middle size in the sense that their y-values fall between the other y-values. They need not be the middle two values according to the

Exhibit 6–3   Thirty Days of Cow Temperatures

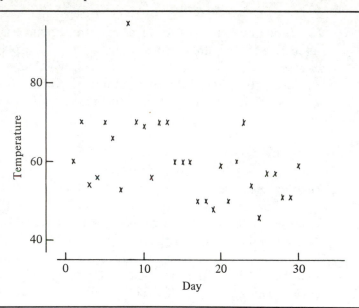

order defined by t—indeed, they need not even be consecutive data points in the sequence.

When using even-length running medians, we must average the t-values as well. The median of an odd-length segment of a data sequence is naturally recorded at the middle t-value of the segment. The natural center of an even-length segment is not at a t-value, so we record the median in the gap between the two middle values of t. A pair of medians then flanks each original t-value. We can align a new y-value with an original t-value by averaging the running medians on either side. We might picture the operation like this:

data values	...	y_5	y_6	y_7	y_8	y_9	y_{10} ...
smoothed by 4's	...	$z_{4.5}$	$z_{5.5}$	$z_{6.5}$	$z_{7.5}$	$z_{8.5}$	$z_{9.5}$ ...
recentered by pairs	...	z_5	z_6	z_7	z_8	z_9	z_{10} ...

Once again we postpone a detailed treatment of the ends of the sequence until Section 6.4.

Of course, the recentering step is just a running median of two because the median of two numbers is also their average. Algebraically, a running median of four, recentered with a running median of two, replaces the data

Exhibit 6–4   Smoothing Cow Temperatures by Running Medians of Three and Five

Day	Temperature (chirps/5 min. − 800)	Smoothed by Running Medians of Three	Smoothed by Running Medians of Five
1	60	60.0	60.0
2	70	60.0	60.0
3	54	56.0	60.0
4	56	56.0	66.0
5	70	66.0	56.0
6	66	66.0	66.0
7	53	66.0	70.0
8	95	70.0	69.0
9	70	70.0	69.0
10	69	69.0	70.0
11	56	69.0	70.0
12	70	70.0	69.0
13	70	70.0	60.0
14	60	60.0	60.0
15	60	60.0	60.0
16	60	60.0	60.0
17	50	50.0	50.0
18	50	50.0	50.0
19	48	50.0	50.0
20	59	50.0	50.0
21	50	59.0	59.0
22	60	60.0	59.0
23	70	60.0	54.0
24	54	54.0	57.0
25	46	54.0	57.0
26	57	57.0	54.0
27	57	57.0	51.0
28	51	51.0	57.0
29	51	51.0	51.0
30	59	59.0	59.0

Source: Data from Enrique de Alba and David L. Zartman, "Testing Outliers in Time Series: An Application to Remotely Sensed Temperatures in Cattle," Special Paper No. 130, Agricultural Experiment Station, New Mexico State University, 1979. Reprinted by permission.

Exhibit 6–5   Cow Temperatures Smoothed by (a) Running Medians of Three and (b) Running Medians of Five

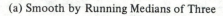

(a) Smooth by Running Medians of Three

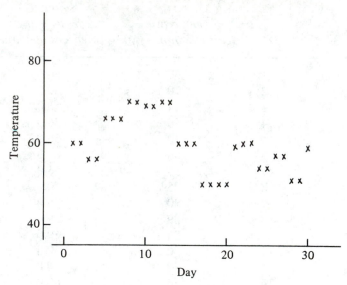

(b) Smooth by Running Medians of Five

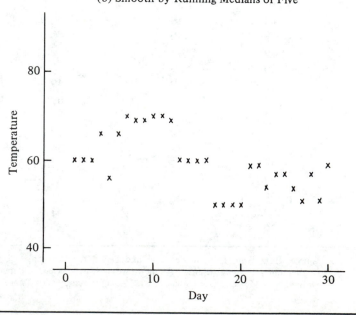

value y_t by

$$z_t = \tfrac{1}{2}(\mathrm{med}\{y_{t-2}, y_{t-1}, y_t, y_{t+1}\} + \mathrm{med}\{y_{t-1}, y_t, y_{t+1}, y_{t+2}\}).$$

This equation uses five data values, y_{t-2} through y_{t+2}, but the first and last values appear in only one of the two segments whose medians are averaged, and thus they have about half the effect of any of the other points. Exhibits 6–6 and 6–7 show results of smoothing the cow temperatures by medians of four and then by medians of two.

span

The number of data values summarized by each median is known as the *span* of the smoother. We have thus far examined smoothers with spans of 2, 3, 4, and 5. Median smoothers with larger spans can resist more outliers. Thus, a span-2 median will be affected by any extraordinary point. Span-3 and span-4 median smoothers will be unaffected by single outliers. A span-3 median will follow an outlying pair, but a span-4 median will cut the size of such a 2-point data spike roughly in half. A span-5 median will be completely resistant to a 2-point spike.

A Shorthand Notation

In order to provide a compact notation for elementary smoothing operations, we refer to them by one-character names. The name for a running median is the single digit corresponding to its span, such as **3** or **5**. When a running median of span 4 is followed by the pair-averaging operation to recenter the results, we use the notation **42**. The two-digit name is appropriate because two operations are involved. (In fact, a few sophisticated combinations insert other elementary operations between a **4** and a **2**.) Since we rarely use running medians of more than 7 points, there is little chance of confusing **42** with a running median of 42 data values. The concatenation of one-character names will be especially convenient in Section 6.3, where we combine elementary smoothing operations in order to gain better performance.

Hanning

running weighted average

We may want a smoothing operation still gentler than **42**. For this we can use a *running weighted average*. It is traditional to smooth data sequences by replacing each data value with the average of the data values around it.

Exhibit 6–6   Smoothing Cow Temperatures by **4** and Then by **2**

Day	Temperature (chirps/5 min. − 800)	Smoothed by 4	Smoothed by 42
1	60	60.0	60.00
2	70	65.0	61.50
3	54	58.0	60.50
4	56	63.0	62.00
5	70	61.0	61.00
6	66	61.0	64.50
7	53	68.0	68.00
8	95	68.0	68.75
9	70	69.5	69.50
10	69	69.5	69.50
11	56	69.5	69.50
12	70	69.5	67.25
13	70	65.0	65.00
14	60	65.0	62.50
15	60	60.0	60.00
16	60	60.0	57.50
17	50	55.0	52.50
18	50	50.0	50.00
19	48	50.0	50.00
20	59	50.0	52.25
21	50	54.5	57.00
22	60	59.5	58.25
23	70	57.0	57.00
24	54	57.0	56.25
25	46	55.5	55.50
26	57	55.5	54.75
27	57	54.0	54.00
28	51	54.0	54.00
29	51	54.0	54.50
30	59	55.0	59.00
		59.0	

Exhibit 6–7   Cow Temperature Smoothed (a) by **4** and (b) by **42**

(a) Smooth by Running Medians of Four

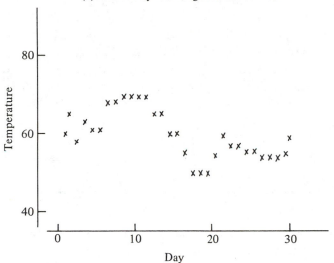

(b) Smooth by Running Medians of Four,
Followed by Running Medians of Two

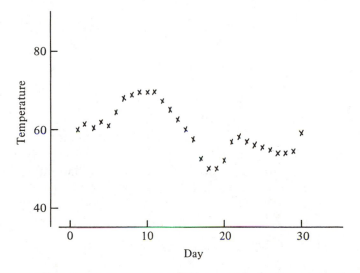

Sometimes the data values are multiplied in each averaging operation by weights. Thus, for example, we might replace y_t by

$$z_t = \tfrac{1}{4}y_{t-1} + \tfrac{1}{2}y_t + \tfrac{1}{4}y_{t+1}.$$

hanning

An unlimited number of running weighted averages are possible (all we require is that the weights—here $\tfrac{1}{4}$, $\tfrac{1}{2}$, $\tfrac{1}{4}$—sum to 1), but we limit ourselves to this particular formula for most data exploration. This smoother is called *hanning*, after Julius von Hann, who advocated its use, and it is denoted by **H**. Any running weighted average will be badly affected by even a single outlier, so we will generally use such smoothers only after outliers have been smoothed away by a running-median smoother.

6.3   Compound Smoothers

While simple running medians will smooth a data sequence and can withstand occasional extraordinary data values, the smooth sequences they produce may describe the data only crudely. We can improve on the description—obtaining data smoothers whose smooth sequences come closer to the data without losing their smoothness—through the judicious combination of smoothing procedures.

Resmoothing

resmoothing

Applying one smoother to the results of a previous smoother is known as *resmoothing*. As with the name **42**, we denote such a series of elementary operations by concatenating their one-character names. If we are working entirely by hand, we may choose to use only running medians of 3 and resmooth repeatedly until further resmoothing yields no further changes. We denote this repeated combination by **3R**.

Reroughing

Running-median smoothers generally smooth a data sequence too much; they remove interesting patterns. A complementary operation can be used to recover smooth patterns from the residuals—that is, from the part called "rough" in the formula

$$\text{data} = \text{smooth} + \text{rough}.$$

We smooth the rough sequence and add the result to the smooth sequence. Our hope is that patterns that have been smoothed away by the first pass of smoothing can be recovered from the rough and used to make the smooth a little more like the original data sequence. By analogy with resmoothing, this operation is called *reroughing*.

reroughing

Exhibits 6–8 and 6–9 show the span-5 median seen in Exhibit 6–4 as reroughed by a span-5 median. We often use the same smoother in both smoothing and reroughing, and we call this using a smoother *twice*. Thus this example illustrates smoothing by **5,twice**.

twicing

Reroughing is an example of an operation found in several exploratory techniques that polish a fit. In the resistant line (Chapter 5), the "reroughing" step involves fitting a line to the residuals and adding this line to the fit. We will see a similar operation in Chapter 8 as the basis for median polish.

4253H

Compound smoothers often combine several elementary smoothers by both resmoothing and reroughing. The early steps in a compound smoother concentrate on protection from outliers in the data sequence. Later steps of resmoothing can then employ a running weighted average. Curiously, running medians of 3 or 5 can alter some rapidly oscillating sequences strangely. For example, the infinite sequence . . . , $+1, -1, +1, -1, +1, -1, . . .$ is not modified at all by a span-5 running median, although the sequence oscillates rapidly. Stranger still, a span-3 running median will invert the sequence, as if each value had been multiplied by -1. Thus, even-span running medians are sometimes preferred—especially when a computer is available to do all the averaging they require.

Similar considerations arise in reroughing because the rough, by design, will contain spikes reflecting the outliers present in the original data and will generally oscillate rapidly. Therefore, the smoothers applied to the rough must also be resistant to these features.

One combination of smoothers that seems to perform quite well is **4253H**. It starts with a running median of four, **4** recentered by **2**. It then resmooths by **5**, by **3**, and finally—now that outliers have been smoothed

Exhibit 6–8   Cow Temperatures, Smoothed by **5** and Reroughed by **5**

Day	Data	Data Smoothed by 5	Rough	Rough Smoothed by 5	5,twice
1	60	60	0	0	60
2	70	60	10	0	60
3	54	60	−6	0	60
4	56	66	−10	0	66
5	70	56	14	−6	50
6	66	66	0	0	66
7	53	70	−17	1	71
8	95	69	26	0	69
9	70	69	1	−1	68
10	69	70	−1	1	71
11	56	70	−14	1	71
12	70	69	1	0	69
13	70	60	10	0	60
14	60	60	0	0	60
15	60	60	0	0	60
16	60	60	0	0	60
17	50	50	0	0	50
18	50	50	0	0	50
19	48	50	−2	0	50
20	59	50	9	0	50
21	50	59	−9	1	60
22	60	59	1	1	60
23	70	54	16	−3	51
24	54	57	−3	1	58
25	46	57	−11	3	60
26	57	54	3	−3	51
27	57	51	6	0	51
28	51	57	−6	0	57
29	51	51	0	0	51
30	59	59	0	0	59

away—by **H**. The result of this smoothing is often reroughed—or polished—by computing residuals, applying the same smoother to them, and adding the result to the smooth of the first pass. This produces the full smoother, **4253H,twice**.

Exhibits 6–10 and 6–11 show an application of this **4253H,twice** step

Exhibit 6–9   Cow Temperatures Smoothed by **5,twice**

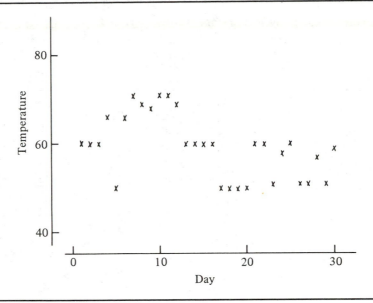

by step. These exhibits make it easy to see how each step affects the data sequence and why we are happy to let the computer do the work. Each column labeled with the name of a smoother shows the result of applying that smoother to the previous column. In Exhibit 6–10, column 7, labeled *Rough 1,* contains the residuals after the first pass of **4253H**, and the succeeding columns smooth these residuals. In Exhibit 6–11, column 13, labeled *Final Smooth,* is the sum of column 6, the first smooth by **4253H**, and column 12, the smoothed **rough**.

6.4   Smoothing the Endpoints

Thus far we have done little to smooth the initial and final values of a data sequence. We cannot smooth these values in the same way as we have smoothed the others because they are not surrounded by enough other values. With a longer-span smoother like **5**, we can forestall the problem by finding shorter-span medians near the endpoints. Thus, for running medians of five,

Exhibit 6–10   Smoothing Cow Temperatures by **4253H**

(1) Temp.	(2) **4**	(3) **2**	(4) **5**	(5) 3(E)*	(6) **H**	(7) *Rough1*
60	60.0	60.00	60.00	60.00	60.0000	0.0000
70	65.0	61.50	60.50	60.50	60.5000	9.5000
54	58.0	60.50	61.00	61.00	61.0000	−7.0000
56	63.0	62.00	61.50	61.50	61.5000	−5.5000
70	61.0	61.00	62.00	62.00	62.5000	7.5000
66	61.0	64.50	64.50	64.50	64.7500	1.2500
53	68.0	68.00	68.00	68.00	67.3125	−14.3125
95	68.0	68.75	68.75	68.75	68.7500	26.2500
70	69.5	69.50	69.50	69.50	69.3125	0.6875
69	69.5	69.50	69.50	69.50	69.5000	−0.5000
56	69.5	69.50	69.50	69.50	68.9375	−12.9375
70	69.5	67.25	67.25	67.25	67.2500	2.7500
70	65.0	65.00	65.00	65.00	64.9375	5.0625
60	65.0	62.50	62.50	62.50	62.5000	−2.5000
60	60.0	60.00	60.00	60.00	60.0000	0.0000
60	60.0	57.50	57.50	57.50	56.8750	3.1250
50	55.0	52.50	52.50	52.50	53.6875	−3.6875
50	50.0	50.00	52.25	52.25	52.3125	−2.3125
48	50.0	50.00	52.25	52.25	52.2500	−4.2500
59	50.0	52.25	52.25	52.25	53.4375	5.5625
50	54.5	57.00	57.00	57.00	55.8125	−5.8125
60	59.5	58.25	57.00	57.00	57.0000	3.0000
70	57.0	57.00	57.00	57.00	56.8125	13.1875
54	57.0	56.25	56.25	56.25	56.2500	−2.2500
46	55.5	55.50	55.50	55.50	55.5000	−9.5000
57	55.5	54.75	54.75	54.75	54.8750	2.1250
57	54.0	54.00	54.50	54.50	54.5625	2.4375
51	54.0	54.00	54.50	54.50	54.5000	−3.5000
51	54.0	54.50	54.50	54.50	54.5000	−3.5000
59	55.0	59.00	59.00	54.50	54.5000	4.5000
	59.0					

*E denotes the endpoint adjustment (Section 6.4).

Exhibit 6–11   Reroughing of Cow Temperatures by **4253H**

(8) **4**	*(9)* **2**	*(10)* **5**	*(11)* **3(E)**	*(12)* **H**	*(13)* *Final* *Smooth*
0.00000	0.00000	0.00000	0.00000	0.00000	60.00000
4.75000	1.00000	0.00000	0.00000	− 0.14063	60.35937
− 2.75000	− 0.87500	− 0.56250	− 0.56250	− 0.42188	60.57812
1.00000	− 0.56250	− 0.56250	− 0.56250	− 0.56250	60.93750
− 2.12500	− 2.12500	− 0.56250	− 0.56250	− 0.28906	62.21093
− 2.12500	1.12500	0.53125	0.53125	0.25781	65.00781
4.37500	2.67188	0.53125	0.53125	0.53125	67.84375
0.96875	0.53125	0.53125	0.53125	0.53125	69.28125
0.09375	0.09375	0.53125	0.53125	0.53125	69.84375
0.09375	0.09375	0.53125	0.53125	0.55078	70.05078
0.09375	0.60938	0.60938	0.60938	0.59375	69.53125
1.12500	0.62500	0.62500	0.62500	0.62109	67.87109
0.12500	0.75000	0.62500	0.62500	0.62500	65.56250
1.37500	1.46875	0.62500	0.62500	0.50781	63.00781
1.56250	0.15625	0.15625	0.15625	− 0.06641	59.93359
− 1.25000	− 1.20313	− 1.20313	− 1.20313	− 1.08203	55.79296
− 1.15625	− 2.07813	− 2.07813	− 2.07813	− 1.85938	51.82812
− 3.00000	− 3.00000	− 2.07813	− 2.07813	− 2.07813	50.23437
− 3.00000	− 3.14063	− 2.07813	− 2.07813	− 2.04688	50.20312
− 3.28125	− 1.95313	− 1.95313	− 1.95313	− 1.40234	52.03515
− 0.62500	1.82813	0.37500	0.37500	− 0.20703	55.60546
4.28125	2.32813	0.37500	0.37500	0.37500	57.37500
0.37500	0.37500	0.37500	0.37500	0.32031	57.13281
0.37500	0.15625	0.15625	0.15625	0.15625	56.40625
− 0.06250	− 0.06250	− 0.06250	− 0.06250	− 0.08594	55.41406
− 0.06250	− 0.37500	− 0.37500	− 0.37500	− 0.29688	54.57812
− 0.68750	− 0.68750	− 0.37500	− 0.37500	− 0.37500	54.18750
− 0.68750	− 0.60938	− 0.37500	− 0.37500	− 0.28516	54.21484
− 0.53125	− 0.01563	− 0.01563	− 0.01563	0.07422	54.57421
0.50000	4.50000	4.50000	0.70313	0.70313	55.20312
4.50000					

we take medians of three for the second and next-to-last values:

$$z_2 = \text{med}\{y_1, y_2, y_3\}$$

$$z_{n-1} = \text{med}\{y_{n-2}, y_{n-1}, y_n\}.$$

The end values, z_1 and z_n, require a different approach. We have thus far been content just to "copy-on"—that is, to use the end values without changing them. We can do better than this by extrapolating from the smoothed values near the end. We first estimate what the next value past the end value might have been. We can't use the end value itself in this estimate because we haven't smoothed it yet. A good, simple approach is to find the straight line that passes through the second and third smoothed values from the end and to place our estimated point on this line at the t-value it would have occupied (see Exhibit 6–12). For equally spaced data with t-spacing Δt, the line at the low end has slope

$$\frac{z_3 - z_2}{\Delta t}$$

We are extrapolating two t-intervals beyond z_2, so the estimated value is

$$\hat{y}_0 = z_2 - 2\Delta t(z_3 - z_2)/\Delta t$$

$$= 3z_2 - 2z_3$$

where the z's are the already smoothed values. Similarly, for the final point we

Exhibit 6–12   The Endpoint Extrapolation

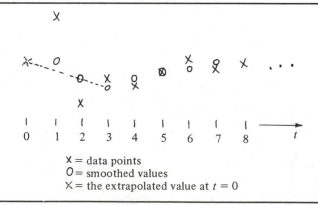

X = data points
O = smoothed values
X = the extrapolated value at $t = 0$

estimate the succeeding point as

$$\hat{y}_{n+1} = 3z_{n-1} - 2z_{n-2}.$$

We then find the median of the extrapolated point, the observed endpoint, and the smoothed point next to the end:

$$z_1 = \text{med}\{\hat{y}_0, y_1, z_2\}$$

$$z_n = \text{med}\{\hat{y}_{n+1}, y_n, z_{n-1}\}.$$

We will not bother with this adjustment every time, but we will usually want to make it at least once at a late step in a compound smoothing. Thus, if we denote this operation by **E**, we might use **4253EH,twice.**

The smoother **42** has an additional end-value problem because it needs to recenter the result of the first smoothing. When we smooth by running medians of four, we obtain a sequence one point longer than the original data sequence. We might denote this longer sequence by $z_{1/2}, z_{1-1/2}, \ldots, z_{n+1/2}$. Here the end values have been copied: $z_{1/2} = y_1, z_{n+1/2} = y_n$. The next values in from each end are medians of two: $z_{1-1/2} = \text{med}\{y_1, y_2\}, z_{n-1/2} = \text{med}\{y_{n-1}, y_n\}$. The subsequent recentering by running medians of 2 restores the sequence to its original length. Again, end values are copied: $z_1 = z_{1/2} \ (=y_1), z_n = z_{n+1/2} \ (=y_n)$. All other values are averages of adjacent values; for example, $z_2 = \text{med}\{z_{1-1/2}, z_{2-1/2}\} = (z_{1-1/2} + z_{2-1/2})/2$.

6.5   Splitting and 3RSSH

When we smooth by hand, we may prefer compound smoothers, such as the repeated running median **3R**, that require fewer calculations. Unfortunately, **3R** has a tendency to chop off peaks and valleys and to leave flat "mesas" and "dales" two points long. We use the special splitting operation named **S** at each 2-point mesa and dale to improve the smooth sequence. We split the data into three pieces—a two-point flat segment, the smooth data sequence to the left of the two points, and the smooth sequence to their right. We then estimate where either point in the flat segment ought to be by referring to the smooth sequence on its own side.

The estimation method is much the same as the endpoint rule discussed in Section 6.4. If, in the smooth by **3R**, the sequence

$$y_{f+1}, y_{f+2}, y_{f+3}, \ldots$$

is to the right of the two-point flat segment

$$y_{f-1}, y_f,$$

we predict what y_{f-1} would have been if it were on the straight line formed by y_{f+1} and y_{f+2}. As we found in extrapolating for the endpoints, we can predict y_{f-1} as

$$3y_{f+1} - 2y_{f+2}.$$

We now use this extrapolated value in a span-3 median centered at y_f:

$$z_f = \text{med}\{3y_{f+1} - 2y_{f+2}, y_f, y_{f+1}\}.$$

Note that all of the values in this operation have already been smoothed by **3R.** This is the only difference between this operation and the endpoint smoothing operation, which uses both the *unsmoothed* end value and nearby smoothed values.

We perform the corresponding operation on the other half of the two-point flat segment; that is, we predict y_f from the line through y_{f-3} and y_{f-2} and use the predicted value in a span-3 median to calculate z_{f-1}. After splitting each two-point mesa and dale, we resmooth the entire sequence by **3R.** Although splitting is tedious by hand, we are likely to need it at only a few places in a data sequence.

One good combination of these operations for smoothing by hand repeats **S** (each time automatically followed by **3R**). It is **3RSSH,twice**. Although it is primarily a hand smoothing technique, the computer programs in this chapter provide **3RSSH,twice** as an option. Exhibit 6–13 shows the steps of **3RSSH** applied to the cow temperatures of Exhibits 6–3 through 6–11.

6.6   Looking at the Rough

We are often as interested in the residual, or rough, sequence as we are in the smooth. The rough can reveal outliers, as well as portions of the sequence that

Exhibit 6–13   Smoothing Cow Temperatures by **3RSSH**

Temp.	3R	S	(3R)	S	(3R)	H
60	60	60	60	60	60	60
70	60	60	60	60	60	60
54	56	**60**	60	60	60	61.5
56	56	**66**	66	66	66	64.5
70	66	66	66	66	66	66
66	66	66	66	66	66	66
53	66	66	66	66	66	66
95	70	**66**	66	66	66	66.75
70	70	**69**	69	69	69	68.5
69	69	**70**	70	**70**	70	69.75
56	69	**70**	70	**70**	70	69.75
70	70	**69**	69	69	69	67
70	70	**60**	60	60	60	62.25
60	60	60	60	60	60	60
60	60	60	60	60	60	60
60	60	60	60	60	60	57.5
50	50	50	50	50	50	52.5
50	50	50	50	50	50	50
48	50	50	50	50	50	50
59	50	50	50	50	50	52.25
50	59	59	59	59	59	56.75
60	60	**60**	59	59	59	59
70	60	**54**	59*	59	59	58.5
54	54	**60**	57	57	57	57.5
46	54	**57**	57	57	57	56.25
57	57	**54**	54	54	54	54
57	57	**51**	51	51	51	51.75
51	51	51	51	51	51	51
51	51	51	51	51	51	51
59	51	51	51	51	51	51

Note: Only the boldface entries are affected by the smoothing operations for that column.
*This value requires two passes of **3**.

seem to be subject to larger fluctuations. We illustrate this by smoothing a sequence of birthrate data.

Exhibit 6–14 shows the number of live births per 10,000 23-year-old women in the United States between 1917 and 1975 (from the data in Exhibit 4–1) and the smooth of that data by **4253H,twice**. The large-scale trends in

Exhibit 6–14 U.S. Birthrate for 23-Year-Old Women, 1917–1975, and Smooth by **4253H,twice**

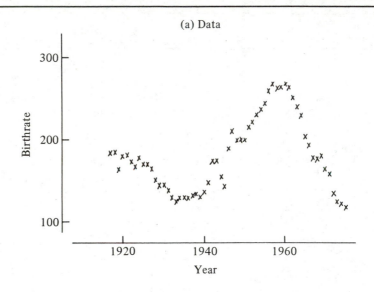

(a) Data

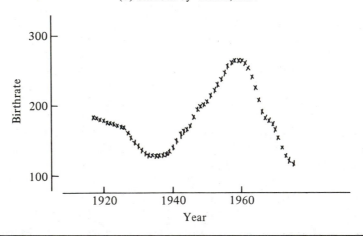

(b) Smooth by **4253H,twice**

birthrate—dropping through the Depression, rising from World War II, and falling again after 1960—are clearly seen in the plot and are well known. The rough sequence, shown in Exhibit 6–15, is more interesting. Birthrates were unstable in the early 1920s, erratic during World War II, and unstable in the 1960s. At other times they have changed rather smoothly.

Exhibit 6–15   Rough of Birthrate

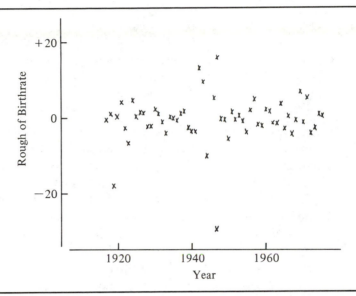

6.7   Smoothing and the Computer

Data smoothing is one of the more tedious EDA techniques to apply by hand. This, combined with the improved performance of the slightly more difficult smoothing methods, makes it a good technique to implement on the computer. The programs in this chapter provide the building blocks of an unlimited variety of data smoothers, but only two compound smoothers, **4253H,twice** and **3RSSH,twice**, are assembled. Other compound smoothers can be constructed with a slight programming effort. (The details will depend on the computer system used.) The compound smoothers provided here perform well in a wide variety of applications and should be sufficient for most needs. If you wish to experiment with other combinations, you should read some of the technical references cited at the end of this chapter. They warn of some of the pitfalls in constructing data smoothers from running medians and provide some guidance.

To use one of the compound smoothers provided here, we need to specify only the data sequence to be smoothed (the data values are assumed to be in sequence order) and where the smooth and rough sequences should be placed. The choice of smoother is the only option.

† 6.8   Algorithms

Data smoothers are often constructed from several similar smoothing operations. The programs for data smoothing take advantage of the great similarity among elementary smoothing operations. These programs, more than any others in this book, are built of many smaller units. This structure makes it easy to build compound smoothers with them.

Several individual algorithms are needed. The most general and most complex is the running-median algorithm. This algorithm uses two temporary work arrays. One of these arrays keeps a "snapshot" of the region of the data sequence surrounding the point to be smoothed. The size of this region is specified by the span of the smoother. The data values are preserved in this work area because each data value participates in the calculation of the smooth values of its successors. Once a data value has been smoothed, its *unsmoothed* value must be remembered for the subsequent smoothing calculations. The second work array holds the same local data values, but they are sorted in order so that the median can be found.

These work arrays are (conceptually) slid along the data sequence so that they can hold the succession of local regions of the data used in the median operations. The smooth value at the current data point is found as the median of the sorted work array. To compute the smooth value at the next data point, the "earliest" data value is found at the beginning of the snapshot work array. (The corresponding value in the data array has already been replaced by its smooth value.) A matching value is then found by searching the sorted work array. (If more than one of the local values is identical to the earliest value, it doesn't matter which is found.) Both the earliest value and its match in the sorted array are removed, and the next data value to be considered as the local region slides along by one *t*-value is then placed in each work array. The sorted work array is re-sorted to find the new median, which is the next smooth value.

The running-median program does not smooth at all near the endpoints. Values not accompanied by at least (span − 1)/2 data values on each side are left unmodified and must be dealt with separately.

Running medians of three are not computed with the same algorithm. Instead, a special program computes them. The program simply compares the three numbers to determine the median. In addition, it reports whether the smooth value is the middle value according to the sequence ordering on *t*. This information makes it easy to check the stopping condition of **3R**.

The algorithms for hanning, smoothing endpoints, and splitting have been specified in Sections 6.2, 6.4, and 6.5, respectively. They are implemented as described in those sections.

Subroutines for each smoothing unit are provided. Each one smooths values near the end explicitly and calls the appropriate general smoothing subroutine.

FORTRAN

The FORTRAN program for data smoothing consists of 13 subroutines: RSM, S4253H, S3RSSH, S2, S3, S4, S5,HANN, S3R, ENDPTS, SPLIT, MEDOF3, and RUNMED. To smooth a data sequence in Y( ), use the FORTRAN statement

CALL RSM(Y, N, SMOOTH, ROUGH, VERSN, ERR)

where

Y( )	is the N-long data vector holding the sequence to be smoothed;
N	is the length of the sequence;
SMOOTH( )	is an N-long array in which the smooth is returned;
ROUGH( )	is an N-long array in which the rough is returned;
VERSN	is a flag = 1 to smooth by **3RSSH,twice,** = 2 to smooth by **4253H,twice;**
ERR	is the error flag, whose values are

    0   normal

    61    $N < 7$—sequence too short to smooth

    62   insufficient work array room—span of running median is greater than allocated space

    63   internal error—possibly an error in the sort program—especially if another sort program has been substituted for the one provided. If so, this could result from incorrect use of that program.

BASIC

The BASIC program for data smoothing consists of 13 subroutines divided in the same way as the FORTRAN subroutines just named. The data sequence of N values to be smoothed is in Y( ). The smooth sequence is returned in Y( ) and the rough sequence in R( ). Arrays C( ) and W( ) are used as work arrays. The array X( ) is not changed because it is likely to be a useful *x*-axis for plotting the smooth and the rough. The version number V1 selects the smoother: V1 = 1 for **3RSSH,twice**, V1 = 2 for **4253H,twice**. Programmers should pay special attention to the use of variables S0 through S9 to save temporary copies of end values.

The smoothing routines require the defined functions and the sorting subroutines. They can nest subroutine calls five deep and use defined functions from the deeper levels. This may strain the capacity of some very small computers.

References

de Alba, Enrique, and David L. Zartman. 1979. "Testing Outliers in Time Series: An Application to Remotely Sensed Temperatures in Cattle," Special Paper No. 130. Agricultural Experiment Station, New Mexico State University, Las Cruces.

Mallows, C.L. 1980. "Some Theory of Nonlinear Smoothers," *Annals of Statistics* 8:695–715.

Velleman, Paul F. 1980. "Definition and Comparison of Robust Nonlinear Data Smoothing Algorithms, *Journal of the American Statistical Association* 75:609–615.

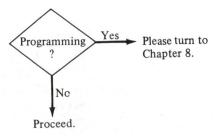

```
5000 REM SMOOTH Y() BY 4253H,TWICE OR 3RSSH,TWICE
5010 REM ENTERED WITH Y() A DATA SEQUENCE IN ORDER (USUALLY
5020 REM ASSUMED TO BE SORTED ON X() WHERE X() EXISTS, BUT NO
5030 REM SORT IS PERFORMED HERE.
5040 REM V1=1 FOR 3RSSH,TWICE:  V1=2 FOR 4253H,TWICE: V1<0 TO ASK.
5050 REM USES C() AND W() FOR TEMPORARY STORAGE AND WORKSPACE
5060 REM RETURNS SMOOTH IN Y(), AND ROUGH IN R(), DOESNT CHANGE X().

5070 IF V1 > 0 THEN 5110
5080 PRINT TAB(M0);"SMOOTHER VERSION: 1=3RSSH,TWICE, 2=4253H,TWICE";
5090 INPUT V1
5100 GO TO 5070
5110 IF N > 6 THEN 5140
5120 PRINT TAB(M0);N;" DATA POINTS IS TOO FEW TO SMOOTH"
5130 RETURN
5140 FOR I = 1 TO N
5150    LET C(I) = Y(I)
5160    LET R(I) = Y(I)
5170 NEXT I
5180 IF V1 > 1 THEN 5230

5190 REM   3RSSH

5200 GOSUB 5520
5210 GO TO 5250

5220 REM   4253H

5230 GOSUB 5420

5240 REM   TWICE (FOR EITHER)

5250 FOR I = 1 TO N
5260    LET X1 = C(I) - Y(I)
5270    LET C(I) = Y(I)
5280    LET Y(I) = X1
5290 NEXT I
5300 IF V1 > 1 THEN 5350

5310 REM   3RSSH

5320 GOSUB 5520
5330 GO TO 5360

5340 REM   4253H

5350 GOSUB 5420
```

```
5360 REM  TWICE
5370 FOR I = 1 TO N
5380    LET Y(I) = Y(I) + C(I)
5390    LET R(I) = R(I) - Y(I)
5400 NEXT I
5410 RETURN

5420 REM SUBROUTINE FOR  4253H
5430 REM OTHER SMOOTHERS CAN BE CONSTRUCTED EASILY BY CALLING THESE
5440 REM SUBROUTINES IN ANOTHER ORDER.

5450 GOSUB 5620
5460 GOSUB 5700
5470 GOSUB 5760
5480 GOSUB 6570
5490 GOSUB 6020
5500 GOSUB 5940
5510 RETURN

5520 REM   SUBROUTINE FOR 3RSSH

5530 GOSUB 6710

5540 REM S8=0 ON EXIT FROM 3R, NOW DO S

5550 GOSUB 6780

5560 REM IF NO CHANGE, THEN DONE

5570 IF S8 = 0 THEN 5610
5580 GOSUB 6710
5590 GOSUB 6780
5600 GOSUB 5940
5610 RETURN

5620 REM 4:S4 IS KEPT FOR 2 LATER,Y(1) ISNT CHANGED--RESULTS IN
                                                  Y(2)-Y(N)

5630 LET S4 = Y(N)
5640 LET S1 = Y(N - 1)
5650 LET S9 = 4
5660 GOSUB 6230
5670 LET Y(2) = (Y(1) + Y(2)) / 2
5680 LET Y(N) = (S1 + S4) / 2
5690 RETURN

5700 REM   2

5710 FOR I = 2 TO N - 1
5720    LET Y(I) = (Y(I) + Y(I + 1)) / 2
5730 NEXT I
5740 LET Y(N) = S4
5750 RETURN
```

```
5760 REM   5

5770 LET S0 = Y(3)
5780 LET S1 = Y(N - 2)
5790 LET S9 = 5
5800 GOSUB 6230

5810 REM   MEDS OF 3 ON ENDS

5820 LET Y1 = Y(1)
5830 LET Y2 = Y(2)
5840 LET Y3 = S0
5850 GOSUB 6140
5860 LET Y(2) = Y2

5870 REM   NOW HIGH END

5880 LET Y1 = S1
5890 LET Y2 = Y(N - 1)
5900 LET Y3 = Y(N)
5910 GOSUB 6140
5920 LET Y(N - 1) = Y2
5930 RETURN

5940 REM   HANN

5950 LET S0 = Y(1)
5960 FOR I = 2 TO N - 1
5970    LET S1 = Y(I)
5980    LET Y(I) = (S0 + Y(I + 1)) / 4 + Y(I) / 2
5990    LET S0 = S1
6000 NEXT I
6010 RETURN

6020 REM   APPLY ENDPOINT RULE TO BOTH ENDS OF Y()

6030 LET Y1 = 3 * Y(2) - 2 * Y(3)
6040 LET Y2 = Y(1)
6050 LET Y3 = Y(2)
6060 GOSUB 6140
6070 LET Y(1) = Y2
6080 LET Y1 = 3 * Y(N - 1) - 2 * Y(N - 2)
6090 LET Y2 = Y(N - 1)
6100 LET Y3 = Y(N)
6110 GOSUB 6140
6120 LET Y(N) = Y2
6130 RETURN
```

```
6140 REM MEDIAN OF Y1,Y2,Y3 RETURNED IN Y2

6150 IF (Y2 - Y1) * (Y3 - Y2) >= 0 THEN 6220

6160 REM Y2 ISNT MEDIAN, COUNT CHANGES. S8 IS CHANGE FLAG.

6170 LET S8 = S8 + 1
6180 IF (Y3 - Y1) * (Y3 - Y2) > 0 THEN 6210
6190 LET Y2 = Y3
6200 GO TO 6220
6210 LET Y2 = Y1
6220 RETURN

6230 REM RUNNING MEDIAN OF LENGTH S9--NO END POINT ROUTINES
6240 REM   S2=POINTER FOR ROTATING SAVE ARRAY,S7=POINTER TO NEXT NUMBER
6250 REM   S3 POINTS TO WHERE THE RESULT GOES.
6260 REM   SORTS IN Y USING W() FOR TEMPORARY STORAGE.

6270 FOR I = 1 TO S9
6280    LET W(I) = Y(I)
6290    LET W(S9 + I) = Y(I)
6300 NEXT I
6310 LET S2 = S9 + 1
6320 LET S3 = FNI((S9 + 2) / 2)
6330 LET S5 = S2 / 2
6340 LET N9 = N
6350 LET N = S9

6360 REM   MAIN LOOP

6370 FOR S7 = S9 + 1 TO N9
6380    GOSUB 1000
6390    LET Y(S3) = FNM(S5)
6400    LET W1 = W(S2)
6410    FOR I = 1 TO S9
6420      IF W(I) = W1 THEN 6460
6430    NEXT I
6440    PRINT "SM ERROR"
6450    STOP
6460    LET W(I) = Y(S7)
6470    LET W(S2) = Y(S7)
6480    LET S2 = S2 + 1
6490    IF S2 <= 2 * S9 THEN 6510
6500    LET S2 = S9 + 1
6510    LET S3 = S3 + 1
6520 NEXT S7
6530 GOSUB 1000
6540 LET Y(S3) = FNM(S5)
6550 LET N = N9
6560 RETURN
```

```
6570 REM SUBROUTINE FOR RUNNING MEDIAN OF LENGTH 3.
6580 REM   THIS IS FASTER  THAN USING THE ABOVE ROUTINE FOR THIS SPECIAL
6590 REM   CASE, AND MAKES 3R EASIER.

6600 LET Y0 = Y(1)
6610 FOR I = 2 TO N - 1
6620    LET Y1 = Y0
6630    LET Y2 = Y(I)
6640    LET Y3 = Y(I + 1)

6650    REM   FIND MEDIAN OF Y1,Y2,Y3--S8 WILL BE S8+1 IF CHANGE IS MADE

6660    GOSUB 6140
6670    LET Y0 = Y(I)
6680    LET Y(I) = Y2
6690 NEXT I
6700 RETURN

6710 REM   SUBROUTINE FOR 3R. REPEAT  3  UNTIL NO CHANGE TAKES PLACE.

6720 LET S8 = 0
6730 GOSUB 6570
6740 IF S8 > 0 THEN 6720

6750 REM   ABOVE LOOP MUST END. NOW DO ENDPOINTS

6760 GOSUB 6020
6770 RETURN

6780 REM   SPLIT 2-PLATEAUS
6790 REM   LOCATE PLATEAUS OF LENGTH 2 AND APPLY ENDPOINT RULES
6800 REM   IF S8=0 ON ENTRY, S8=0 ON EXIT IFF NO CHANGES MADE
6810 REM THIS ROUTINE USES W(1)-W(6) AS TEMPORARY STORAGE.
6820 REM   A SLIDING WINDOW ON Y().

6830 LET N2 = N - 2

6840 REM   INITIALIZE WITH FIRST 4 POINTS

6850 FOR I = 1 TO 4
6860    LET W(I + 2) = Y(I)
6870 NEXT I

6880 REM   Y(1) AND Y(2) ARE A PLATEAU IF OK ON RIGHT--FAKE THE LEFT

6890 LET W(2) = Y(3)

6900 REM   I1 IS POINTER FOR Y()

6910 LET I1 = 1
```

```
6920 REM   HUNT FOR 2-PLATEAUS

6930 IF W(3) <> W(4) THEN 7100
6940 IF (W(3) - W(2)) * (W(5) - W(4)) >= 0 THEN 7100

6950 REM   W(3)&W(4) (=Y(I1)&Y(I1+1)) ARE A PLATEAU
6960 REM   APPLY RIGHT ENDPOINT RULE AT I1, IF WE CAN

6970 IF I1 < 3 THEN 7040
6980 LET Y1 = 3 * W(2) - 2 * W(1)
6990 LET Y2 = W(3)
7000 LET Y3 = W(2)
7010 GOSUB 6140
7020 LET Y(I1) = Y2

7030 REM   APPLY LEFT END POINT RULE AT I1+1 IF WE CAN

7040 IF I1 >= N2 THEN 7100
7050 LET Y1 = 3 * W(5) - 2 * W(6)
7060 LET Y2 = W(4)
7070 LET Y3 = W(5)
7080 GOSUB 6140
7090 LET Y(I1 + 1) = Y2

7100 REM   SLIDE THE WINDOW

7110 FOR I = 1 TO 5
7120    LET W(I) = W(I + 1)
7130 NEXT I
7140 LET I1 = I1 + 1
7150 IF I1 >= N2 THEN 7180
7160 LET W(6) = Y(I1 + 3)
7170 GO TO 6920

7180 REM   LAST 2 POINTS ARE A PLATEAU IF OK ON LEFT--FAKE THE RIGHT

7190 LET W(6) = W(3)
7200 IF I1 < N THEN 6920
7210 RETURN
```

```
      SUBROUTINE RSM(Y, N, SMOOTH, ROUGH, VERSN, ERR)
C
      INTEGER N, VERSN, ERR
      REAL Y(N), SMOOTH(N), ROUGH(N)
C  MAIN PROGRAM FOR NONLINEAR SMOOTHERS.
C
C  ON ENTRY:
C  Y()  IS A DATA SEQUENCE OF  N  VALUES
C  VERSN  SPECIFIES THE SMOOTHER TO BE USED
C     VERSN=1 SPECIFIES 3RSSH, TWICE
C     VERSN=2 SPECIFIES 4253H, TWICE
C  ON EXIT:
C  SMOOTH()  AND  ROUGH()  CONTAIN THE SMOOTH AND ROUGH RESULTING FROM
C  THE SMOOTHING OPERATION.  NOTE THAT
C       Y(I) = SMOOTH(I) + ROUGH(I)
C  FOR EACH I FROM 1 TO N.
C
C  LOCAL VARIABLE
C
      INTEGER I
C
C
      IF (N .GT. 6) GO TO 10
      ERR = 61
      GO TO 999
   10 DO 20 I = 1, N
         SMOOTH(I) = Y(I)
   20 CONTINUE
      IF (VERSN .EQ. 1) CALL S3RSSH(SMOOTH, N, ERR)
      IF (VERSN .EQ. 2) CALL S4253H(SMOOTH, N, ERR)
      IF (ERR .NE. 0) GO TO 999
C
C  COMPUTE ROUGH FROM FIRST SMOOTHING
C
      DO 30 I = 1, N
         ROUGH(I) = Y(I) - SMOOTH(I)
   30 CONTINUE
C
C  REROUGH SMOOTHERS ("TWICING")
C
      IF (VERSN .EQ. 1) CALL S3RSSH(ROUGH, N, ERR)
      IF (VERSN .EQ. 2) CALL S4253H(ROUGH, N, ERR)
      IF (ERR .NE. 0 ) GO TO 999
      DO 40 I = 1, N
        SMOOTH(I) = SMOOTH(I) + ROUGH(I)
        ROUGH(I) = Y(I) - SMOOTH(I)
   40 CONTINUE
  999 RETURN
      END
```

191

```
      SUBROUTINE S3RSSH(Y, N, ERR)
C
C SMOOTH Y() BY 3RSSH, TWICE
C
      INTEGER N, ERR
      REAL Y(N)
C
C   LOCAL VARIABLE
C
      LOGICAL CHANGE
C
      CALL S3R(Y, N)
      CHANGE = .FALSE.
      CALL SPLIT(Y, N, CHANGE)
      IF (.NOT. CHANGE) GO TO 10
      CALL S3R(Y, N)
      CHANGE = .FALSE.
      CALL SPLIT(Y, N, CHANGE)
      IF (CHANGE) CALL S3R(Y, N)
   10 CALL HANN(Y, N)
  999 RETURN
      END


      SUBROUTINE S4253H(Y, N, ERR)
C
C   SMOOTH BY 4253H
C
      INTEGER N, ERR
      REAL Y(N)
C
C   LOCAL VARIABLES
C
      REAL ENDSAV, WORK(5), SAVE(5)
      INTEGER NW
      LOGICAL CHANGE
      DATA NW/5/
C
      CHANGE =.FALSE.
C
      CALL S4(Y, N, ENDSAV, WORK, SAVE, NW, ERR)
      IF(ERR .EQ. 0) CALL S2(Y, N, ENDSAV)
      IF(ERR .EQ. 0) CALL S5(Y, N, WORK, SAVE, NW, ERR)
      IF(ERR .EQ. 0) CALL S3(Y, N, CHANGE)
      IF(ERR .EQ. 0) CALL ENDPTS(Y, N)
      IF(ERR .EQ. 0) CALL HANN(Y, N)
  999 RETURN
      END
```

```
      SUBROUTINE S4(Y, N, ENDSAV, WORK, SAVE, NW, ERR)
C
C   SMOOTH BY RUNNING MEDIANS OF 4.
C
      INTEGER N, NW, ERR
      REAL Y(N), ENDSAV, WORK(NW), SAVE(NW)
C
C   LOCAL VARIABLES
C
      REAL ENDM1, TWO
      DATA TWO/2.0/
C
C   EVEN LENGTH MEDIANS OFFSET THE OUTPUT SEQUENCE TO THE HIGH END,
C    SINCE THEY CANNOT BE SYMMETRIC.  ENDSAV IS LEFT HOLDING Y(N) SINCE
C    THERE IS NO OTHER ROOM FOR IT.  Y(1) IS UNCHANGED.
C
      ENDSAV = Y(N)
      ENDM1 = Y(N-1)
      CALL RUNMED(Y, N, 4, WORK, SAVE, NW, ERP)
C
      Y(2) = (Y(1) + Y(2))/TWO
      Y(N) = (ENDM1 + ENDSAV)/TWO
  999 RETURN
      END

      SUBROUTINE S2(Y, N, ENDSAV)
C
C   SMOOTH BY RUNNING MEDIANS (MEANS) OF 2.
C   USED TO RECENTER RESULTS OF RUNNING MEDIANS OF 4.
C    ENDSAV HOLDS THE ORIGINAL Y(N).
C
      INTEGER N
      REAL Y(N), ENDSAV
C
C   LOCAL VARIABLES
C
      INTEGER NM1,  I
      REAL TWO
      DATA TWO/2.0/
C
      NM1 = N-1
      DO 10 I = 2, NM1
        Y(I) = (Y(I+1)+Y(I))/TWO
   10 CONTINUE
      Y(N) = ENDSAV
  999 RETURN
      END
```

```
      SUBROUTINE S5(Y, N, WORK, SAVE, NW, ERR)
C
C   SMOOTH BY RUNNING MEDIANS OF 5.
C
      INTEGER N, NW, ERR
      REAL Y(N), WORK(NW), SAVE(NW)
C
C   LOCAL VARIABLES
C
      LOGICAL CHANGE
      REAL YMED1, YMED2
C
      CHANGE = .FALSE.
C
      CALL MEDOF3(Y(1), Y(2), Y(3), YMED1, CHANGE)
      CALL MEDOF3(Y(N), Y(N-1), Y(N-2), YMED2, CHANGE)
      CALL RUNMED(Y, N, 5, WORK, SAVE, NW, ERR)
      Y(2) = YMED1
      Y(N-1) = YMED2
  999 RETURN
      END



      SUBROUTINE HANN(Y, N)
C
C   3-POINT SMOOTH BY MOVING AVERAGES WEIGHTED 1/4,  1/2,  1/4.
C   THIS IS CALLED  HANNING.
C
      INTEGER N
      REAL Y(N)
C
C   LOCAL VARIABLES
C
      INTEGER I, NM1
      REAL Y1, Y2, Y3
C
      NM1 = N-1
      Y2 = Y(1)
      Y3 = Y(2)
C
      DO 10 I = 2, NM1
         Y1 = Y2
         Y2 = Y3
         Y3 = Y(I+1)
         Y(I) = (Y1 + Y2 + Y2 + Y3)/4.0
   10 CONTINUE
  999 RETURN
      END
```

```
      SUBROUTINE S3(Y, N, CHANGE)
C
C  COMPUTE RUNNING MEDIAN OF 3 ON Y().
C  SETS CHANGE  .TRUE.  IF ANY CHANGE IS MADE.
C
      INTEGER N
      REAL Y(N)
      LOGICAL CHANGE
C
C  LOCAL VARIABLES
C
      REAL Y1, Y2, Y3
      INTEGER NM1
C
      Y2=Y(1)
      Y3=Y(2)
      NM1 = N-1
      DO 10 I = 2, NM1
        Y1=Y2
        Y2=Y3
        Y3=Y(I+1)
        CALL MEDOF3(Y1, Y2, Y3, Y(I), CHANGE)
   10 CONTINUE
  999 RETURN
      END


      SUBROUTINE S3R(Y, N)
C
C  COMPUTE REPEATED RUNNING MEDIANS OF 3.
C
      INTEGER N
      REAL Y(N)
C
C  LOCAL VARIABLE
C
      LOGICAL CHANGE
C
   10 CHANGE = .FALSE.
      CALL S3(Y, N, CHANGE)
      IF (CHANGE) GO TO 10
      CALL ENDPTS(Y, N)
  999 RETURN
      END
```

```
      SUBROUTINE MEDOF3(X1, X2, X3, XMED, CHANGE)
C
C  PUT THE MEDIAN OF X1, X2, X3 IN XMED AND
C  SET  CHANGE  .TRUE. IF THE MEDIAN ISNT X2.
C
      REAL X1, X2, X3, XMED
      LOGICAL CHANGE
C
C  LOCAL VARIABLES
C
      REAL Y1, Y2, Y3
C
      Y1=X1
      Y2=X2
      Y3=X3
C
      XMED = Y2
      IF((Y2-Y1) * (Y3-Y2) .GE. 0.0 ) GO TO 999
      CHANGE = .TRUE.
      XMED = Y1
      IF ((Y3-Y1) * (Y3-Y2) .GT. 0.0 ) GO TO 999
      XMED = Y3
  999 RETURN
      END


      SUBROUTINE ENDPTS(Y, N)
C
C  ESTIMATE SMOOTHED VALUES FOR BOTH END POINTS OF THE SEQUENCE IN Y()
C  USING THE END POINT EXTRAPOLATION RULE.
C  ALL THE VALUES IN Y() EXCEPT THE END POINTS HAVE BEEN SMOOTHED.
C
      INTEGER N
      REAL Y(N)
C
C  LOCAL VARIABLES
C
      REAL Y0,  YMED
      LOGICAL CHANGE
C
      CHANGE = .FALSE.
C
C  LEFT END
C
      Y0 = 3.0*Y(2) - 2.0*Y(3)
      CALL MEDOF3(Y0, Y(1), Y(2), YMED, CHANGE)
      Y(1) = YMED
```

```
C
C   RIGHT END
C
      Y0= 3.0*Y(N-1)-2.0*Y(N-2)
      CALL MEDOF3(Y0, Y(N), Y(N-1), YMED, CHANGE)
      Y(N) = YMED
  999 RETURN
      END


      SUBROUTINE SPLIT(Y, N, CHANGE)
C
C   FIND 2-FLATS IN Y() AND APPLY SPLITTING ALGORITHM.
C
      INTEGER N
      REAL Y(N)
      LOGICAL CHANGE
C
C LOCAL VARIABLES
C
      REAL W(6), Y1
      INTEGER I1, I, NM2
C
C   W() IS A WINDOW 6 POINTS WIDE WHICH IS SLID ALONG Y().
C
      NM2 = N-2
      DO 10 I = 1, 4
        W(I+2) = Y(I)
   10 CONTINUE
C
C   IF Y(1)=Y(2) .NE. Y(3), TREAT FIRST 2 LIKE A 2-FLAT WITH END PT RULE
C
      W(2)=Y(3)
      I1 = 1
   20 IF (W(3) .NE. W(4)) GO TO 40
      IF ( (W(3)-W(2)) * (W(5)-W(4)) .GE. 0.0 ) GO TO 40
C   W(3) AND W(4) FORM A 2-FLAT.
      IF ( I1 .LT. 3) GO TO 30
C
C   APPLY RIGHT END PT RULE AT I1
C
      Y1= 3.0 * W(2) - 2.0 * W(1)
      CALL MEDOF3(Y1, W(3), W(2), Y(I1), CHANGE)
   30 IF (I1 .GE. NM2) GO TO 40
C
C   APPLY LEFT END PT RULE AT I1+1
C
      Y1 = 3.0*W(5) - 2.0*W(6)
      CALL MEDOF3(Y1, W(4), W(5), Y(I1+1), CHANGE)
```

```
C
C   SLIDE WINDOW
C
   40 DO 50 I = 1, 5
         W(I) = W(I+1)
   50 CONTINUE
         I1 = I1+1
         IF (I1 .GE. NM2) GO TO 60
         W(6) = Y(I1+3)
         GO TO 20
C
C  APPLY RULE TO LAST 2 POINTS IF NEEDED.
C
   60 W(6)=W(3)
         IF(I1 .LT. N ) GO TO 20
  999 RETURN
         END


         SUBROUTINE RUNMED(Y, N, LEN, WORK, SAVE, NW, ERR)
C  SMOOTH Y() BY RUNNING MEDIANS OF LENGTH LEN.
C  NOTE: USE S3 FOR RUNNING MEDIANS OF 3 INSTEAD OF  RUNMED.
C
         INTEGER N, LEN, NW, ERR
         REAL Y(N), WORK(NW), SAVE(NW)
C
C  FUNCTION
C
         REAL MEDIAN
C
C  LOCAL VARIABLES
C
         REAL TEMP, TWO
         INTEGER SAVEPT,   SMOPT,   LENP1, I, J
C
C  WORK() IS A LOCAL ARRAY IN WHICH DATA VALUES ARE SORTED.
C
         DATA TWO/2.0/
C
C  SAVE() ACTS AS A WINDOW ON THE DATA.
C
         IF(LEN .LE. NW) GO TO 5
         ERR = 62
         GO TO 999
    5    DO 10 I=1, LEN
            WORK(I) = Y(I)
            SAVE(I) = Y(I)
   10 CONTINUE
```

```
      SAVEPT = 1
      SMOPT = INT((FLOAT(LEN) + TWO)/TWO)
      LENP1 = LEN + 1
      DO 50 I = LENP1, N
        CALL SORT(WORK, LEN, ERR)
        IF(ERR .NE. 0) GO TO 999
        Y(SMOPT) = MEDIAN(WORK, LEN)
        TEMP = SAVE(SAVEPT)
        DO 20 J=1, LEN
          IF (WORK(J) .EQ. TEMP ) GO TO 30
  20    CONTINUE
        ERR = 63
      GO TO 999
  30   WORK(J) = Y(I)
        SAVE(SAVEPT) = Y(I)
        SAVEPT = MOD(SAVEPT, LEN)+1
        SMOPT = SMOPT + 1
  50 CONTINUE
      CALL SORT(WORK, LEN, ERR)
      IF(ERR .NE. 0) GO TO 999
      Y(SMOPT) = MEDIAN(WORK, LEN)
 999 RETURN
      END
```

Chapter 7 _____

Coded Tables

two-way
table

We have examined data with several types of structure. In this chapter we consider another data structure, the *table*. Tables of numbers are a common way to organize data when each data value is related simultaneously to two factors. For example, Exhibit 7–1 shows the death rates (in deaths per 1000 for men) reported in a British study of the health effects of smoking. Each row of the table in Exhibit 7–1 reports a different cause of death, and each column holds data for different amounts of smoking. Any number in the table can easily be identified with its row and column labels. Thus, for example, non-smokers died of chronic bronchitis at the rate of about .12 per 1000.

The kinds of patterns we might look for in tables are much the same as those we have sought in other kinds of data, except that in tables we have three things to keep track of: the row identity, the column identity, and the data value in the cell. For example, if, as in Exhibit 7–1, the columns have a natural order, we might look for trends as we move from left to right in the table. These might be an overall trend—for example, men who smoke more die at a greater rate than non-smokers—or trends in single rows—for example, this trend is especially strong for lung cancer. Of course, if the rows had a natural order (say, from top to bottom in the table), we might also look for trends against this order.

Exhibit 7–1 Standardized Death Rates (per 1000) for Men in Various Smoking Classes by Cause of Death

		Smoking Class		
Cause of Death	*None*	*1–14 Grams*	*15–24 Grams*	*25+ Grams*
Cancers				
Lung	0.07	0.47	0.86	1.66
Upper respiratory	0.00	0.13	0.09	0.21
Stomach	0.41	0.36	0.10	0.31
Colon and rectum	0.44	0.54	0.37	0.74
Prostate	0.55	0.26	0.22	0.34
Other	0.64	0.72	0.76	1.02
Respiratory diseases				
Pulmonary TB	0.00	0.16	0.18	0.29
Chronic bronchitis	0.12	0.29	0.39	0.72
Other	0.69	0.55	0.54	0.40
Coronary thrombosis	4.22	4.64	4.60	5.99
Other cardiovascular	2.23	2.15	2.47	2.25
Cerebral hemorrhage	2.01	1.94	1.86	2.33
Peptic ulcer	0.00	0.14	0.16	0.22
Violence	0.42	0.82	0.45	0.90
Other diseases	1.45	1.81	1.47	1.57

Source: J. Berkson, "Smoking and Lung Cancer: Some Observations on Two Recent Reports," *Journal of the American Statistical Association* 53 (1958):28–38. Reprinted by permission.

Note: Rates are not age-adjusted.

In the table in Exhibit 7–1, the rows have no natural order. They merely label categories for different causes of death. We might look for differences among the categories—for example, fewer deaths from peptic ulcer. At a slightly more sophisticated level, we might ask whether the patterns we noted across columns change from row to row. In Exhibit 7–1 we can see that they do. Lung cancer death rates show a strong trend with increased smoking; death rates from "other respiratory" (non-cancerous) diseases show a slight decrease as smoking increases.

Finally, as in every exploratory examination of data, we look for outliers. In Exhibit 7–1 an entire row—coronary thrombosis—is prominent as the overwhelming major cause of death among men, and the cell for heavy smokers in this row is substantially larger than the rest of the row.

7.1   Displaying Tables

coded table

Searching large tables for patterns is often tedious. Instead, we need a display that will tame the clutter of numbers in large tables yet reveal the kinds of patterns that we look for in tables. The structure of a table encourages use of a display that preserves the row-by-column shape. The *coded table* does this job neatly.

In a coded table we replace the data with one-character codes that summarize their behavior. The scheme for assigning codes is much like the one we used to construct boxplots in Chapter 3. Data values are identified as being (1) in the middle 50% of the data, between the hinges (coded with a dot, ·), (2) above or below the hinges but within the fences (coded + or −), (3) outside the inner fences (coded # for "double plus" or = for "double minus"), or (4) far outside (coded P for "PLUS" or M for "MINUS"). If a cell is entirely empty, it is coded with a blank. Exhibit 7–2 shows the result of coding the death rates of Exhibit 7–1. The patterns are now actually clearer because we are no longer trying to read 60 numbers and can concentrate on the patterns.

7.2   Coded Tables from the Computer

Coded tables of moderate size are easy to make by hand. All we need are the hinges and fences, which are easy to find from a letter-value display. It is natural to produce a coded table on the computer when the data are already in the machine, but computer-produced coded tables have some additional advantages. A coded table condenses a large table effectively. Only two spaces are needed for each cell of the table rather than the six or more needed to print the numbers. (If we need to print a bigger table, we can omit the space between coding symbols.)

When we have a table in which both rows and columns are ordered and equally spaced, the coded table can serve as a rough contour plot. The codes are chosen so that more extreme points are darker in order to enhance this interpretation.

The computer allows us to make coded tables for more complicated data than we might ordinarily analyze by hand. Some data tables, especially from designed experiments, can have several numbers in each cell of the table. Exhibit 7–3 shows an example in which test animals were given one of three poisons and treated by one of four treatments. Four animals were assigned to each combination of poison and treatment, and the table reports the number of hours each animal survived. Two coded tables are useful here: a coded table of

Exhibit 7–2 Summaries of the Male Death Rates of Exhibit 7–1, Including a Coded Table

```
STEM-AND-LEAF DISPLAY
UNIT = .1
1   2   REPRESENTS 1.2


12        +0* 000001111111
23          T 22222233333
(10)        F 4444445555
27          S 667777
21         0· 889
18         1* 0
17          T
17          F 445
14          S 6
13         1· 889
10         2* 01
 8          T 223
 5          F 4
```

HI: 42, 46, 46, 59,

LETTER-VALUE DISPLAY

$n = 60$

	Depth	Low	High	Mid	Spread
M	30.5		.54	.54	
H	15.5	.24	1.52	.88	1.28
E	8	.13	2.23	1.18	2.10
D	4.5	.08	3.345	1.7125	3.265
C	2.5	0	4.62	2.31	4.62
B	1.5	0	5.315	2.6575	5.315
	1	0	5.99	2.995	5.99

Exhibit 7–2 (continued)

Coded Table

	None	1–14	15–24	25+
Cancers				
Lung	–	·	·	+
Upper respiratory	–	–	–	–
Stomach	·	·	–	·
Colon and rectum	·	·	·	·
Prostate	·	·	–	·
Other	·	·	·	·
Pulmonary TB	–	–	–	·
Chronic bronchitis	–	·	·	·
Other respiratory	·	·	·	·
Coronary thrombosis	#	#	#	P
Other cardiovascular	+	+	+	+
Cerebral hemorrhage	+	+	+	+
Peptic ulcer	–	–	–	–
Violence	·	·	·	·
Other diseases	·	+	·	+

M	Far outside low
=	Below low inner fence (outside)
–	Below lower hinge but within inner fence
·	Between hinges
+	Above upper hinge but within inner fence
#	Above high inner fence (outside)
P	Far outside high

the lowest value in each cell, and a coded table of the highest value in each cell. For both tables the hinges and fences are determined by the *entire* data set of all $3 \times 4 \times 4 = 48$ numbers, although only 12 numbers are coded. Exhibit 7–4 shows the resulting coded tables. The table of maximum values warns of some possible strays.

A third alternative is useful for displaying residuals from a median polish—a technique explained in the next chapter. In this table we display the most extreme (largest in magnitude) number in each cell to highlight possible outliers.

The coded table programs in this chapter require that tables be represented in three arrays. One array holds the data values, a parallel array holds the corresponding row numbers, and a third and also parallel array holds the corresponding column numbers. Thus the simple table

10	20
30	40

Exhibit 7–3   Survival Times of Each of Four Animals After Administration of One of Three Poisons and One of Four Treatments (unit = 10 hours)

	Treatment			
Poison	A	B	C	D
I	0.31	0.82	0.43	0.45
	0.45	1.10	0.45	0.71
	0.46	0.88	0.63	0.66
	0.43	0.72	0.76	0.62
II	0.36	0.92	0.44	0.56
	0.29	0.61	0.35	1.02
	0.40	0.49	0.31	0.71
	0.23	1.24	0.40	0.38
III	0.22	0.30	0.23	0.30
	0.21	0.37	0.25	0.36
	0.18	0.38	0.24	0.31
	0.23	0.29	0.22	0.33

Source: G.E.P. Box and D.R. Cox, "An Analysis of Transformations," *Journal of the Royal Statistical Society, Series B* 26 (1964):211–243. Reprinted by permission.

Exhibit 7–4   Coded Tables for Exhibit 7–3

Minimum Value in Each Cell

	A	B	C	D
I	·	+	·	·
II	−	·	·	·
III	−	−	−	−

Maximum Value in Each Cell

	A	B	C	D
I	·	+	+	+
II	·	#	·	+
III	−	·	−	·

would be described to the programs as

Data	Row	Column
10	1	1
20	1	2
30	2	1
40	2	2

While this structure uses slightly more space than other ways of storing a table, it offers great flexibility. For example, it easily accommodates an empty cell: The combination of its row and column numbers simply never appears. Similarly, multiple data values in a cell are specified by repeating the cell's row and column numbers for each data value. When the table has more than one data value in some cells, the programs must be told whether to code the maximum, minimum, or most extreme value in the cell.

7.3   Coded Tables and Boxplots

Boxplots and coded tables display data in similar ways. Both describe overall patterns in the data and highlight individual extraordinary data values, and both use letter values as a basis for these descriptions. Therefore, it is not surprising that these two displays complement each other well.

Coded tables preserve the row and column location of each data value. This helps to reveal two-dimensional patterns but can be distracting when we want to make comparisons among rows or columns alone. When that is our goal, boxplots may do better.

Exhibit 7–5 shows a table of the U.S. birthrate (live births per 1000 women aged 15–44 years) recorded monthly from 1937 through 1947, and Exhibit 7–6 shows a coded table for the same data. As we saw when we smoothed annual birthrates in the last chapter (see Exhibits 6–14 and 6–15), this period witnessed rapid changes in the U.S. birthrate due, in part, to World War II. The monthly data allow us to examine these changes more closely.

The coded table in Exhibit 7–6 shows some of the patterns we would expect: lower birthrates in 1937–1940 beginning to increase in the early 1940s, decline in the late years of World War II, and the sharp increase of the postwar baby boom. We can now see that the increases in both 1942 and 1946 accelerated in July and August of those years.

Exhibit 7–5   U.S. Birthrate (live births per 1000 women aged 15–44 years) by Month, 1937–1947

	January	*February*	*March*	*April*	*May*	*June*
1937	75.62	78.36	78.58	75.18	74.90	75.87
1938	79.73	81.62	80.27	77.85	76.80	77.54
1939	77.89	79.22	79.26	76.36	73.75	75.74
1940	77.74	80.48	79.15	77.04	77.44	79.25
1941	80.41	82.82	82.98	81.19	77.52	85.23
1942	86.28	88.60	87.23	83.07	81.97	86.23
1943	99.45	99.59	96.76	92.30	89.63	93.83
1944	88.65	89.69	85.69	82.68	83.37	89.35
1945	87.76	88.14	85.62	82.33	82.21	85.88
1946	81.50	83.56	83.45	83.28	85.22	91.35
1947	123.12	120.83	117.69	109.10	109.53	112.55

	July	*August*	*September*	*October*	*November*	*December*
1937	80.73	83.10	82.06	75.77	73.59	74.26
1938	82.87	83.85	82.76	78.13	75.58	74.14
1939	80.79	82.01	82.21	77.60	73.60	72.19
1940	83.69	85.03	84.69	78.97	76.19	76.23
1941	91.45	89.51	86.72	80.84	80.43	81.43
1942	91.65	95.58	101.85	101.62	97.60	95.68
1943	97.87	98.71	98.12	92.13	87.93	86.31
1944	95.30	94.79	91.99	88.48	88.70	87.58
1945	89.15	89.92	90.30	85.31	83.17	81.94
1946	104.41	113.96	122.52	123.61	124.90	123.21
1947	114.79	115.21	115.44	111.08	107.22	103.93

Source: U.S. Department of Health, Education and Welfare, *Seasonal Variations of Births, U.S. 1933–1963*, National Center for Health Statistics, Series 21, no. 9.

This pattern raises the question of whether birthrates, even in times of rapid change, show a seasonal cycle. To answer this question, we need to compare the columns of the table. Exhibit 7–7 shows the 12 boxplots of the birthrates by month. There is clearly an annual cycle; birthrates are lowest in April and May, are highest in the summer, and seem to cycle smoothly month-to-month. The cycle is clear in the sequence of medians, in both the low and high hinges, and even in the outliers (mostly values from 1946 and 1947).

Exhibit 7–6   Coded Table of Monthly Birthrates 1937–1947 (from Exhibit 7–5)

	J	F	M	A	M	J	J	A	S	O	N	D
1937	−	−	−	−	−	−	.	.	.	−	−	−
1938	.	.	.	−	−	−	.	.	.	−	−	−
1939	−	−	−	−	−	−	.	.	.	−	−	−
1940	−	.	−	−	−	−	.	.	.	−	−	−
1941	.	.	.	.	−	.	.	.	.	.	.	.
1942	.	.	.	.	.	.	.	+	+	+	+	+
1943	+	+	+	+	.	+	+	+	+	.	.	.
1944	.	.	.	.	.	.	+	+	.	.	.	.
1945	.	.	.	.	.	.	.	.	.	.	.	.
1946	.	.	.	.	.	.	+	#	#	#	#	#
1947	#	#	#	+	+	#	#	#	#	+	+	+

We might have had some hints of this annual cycle from the coded table, but we certainly could not see the cycle with this clarity.

It is easy to use the programs in this book to obtain boxplots by rows or columns because tables are specified by separate arrays holding row numbers and column numbers (see Section 7.2). We need only specify that either the column-number array (as in the monthly birthrate example) or the row-number array should be the group-identifying array for the boxplot program (see Section 3.8). For some tables we might want to examine a coded table, boxplots by columns, *and* boxplots by rows. If we want to go further in analyzing the birthrate data, we might unravel the table to form a month-by-month time series and apply the data-smoothing methods of Chapter 6. It would probably be interesting to put the rough sequence back into an 11 by 12 table and look again at the coded table after the year-to-year trend and annual cycle have been removed. The methods of the next chapter provide yet another way to analyze this table.

† 7.4   Algorithms

The coded table programs accept data in the form described in Section 7.2. First, the data array is copied and sorted to find the hinges and fences. Then,

Exhibit 7–7   Boxplots of U.S. Birthrate by Month (from Data in Exhibit 7–5)

```
                              -----------
            Jan.      ---I  +        I-----------                    0
                              -----------

                              ----------
            Feb.       --I  +      I-----------                     0
                              ---------

                              --------
            Mar.       -I  + I-              *                      0
                              --------

                            ------
            Apr.     ---I    +-           *                0
                            ------

                            ----------
            May     ---I     +  I-----                    0
                            ---------

                            -------------
            Jun.     ---I      +    I---                           *
                            -------------

                              ----------------
            Jul.        ---I        +    I-------------------
                              ----------------

                              ----------------
            Aug.       ---I    +      I--------------------
                              ----------------

                              ------------------
            Sep.       --I       +         I--------------------------
                              ------------------

                            ------------------
            Oct.     ---I        +        I----------------------------
                            ------------------

                          --------------------
            Nov.    --I     +        I----------------              *
                          ------------------

                          -----------------
            Dec.    ---I   +        I------------              *
                          ------------------
```

the table is re-structured so that the cell in the top row and leftmost column comes first, followed by the rest of the first-row cells in column-number order, from left to right. If more than one value is found for a cell, the maximum, minimum, or most extreme value is kept, depending on which alternative has been specified. The first-row values are followed by the values in the second row, from left to right, and so on. The resulting array is said to be in row-major format. The programs use an internally determined code to mark any empty cells. This code is not a missing-value code and is not used outside these programs. The empty code is generated internally to ensure that it is unique.

The re-ordered table is now in the right order for generating the coded table. Values are considered in turn. They are compared to the hinges and fences, and their codes are printed a line at a time. Empty cells appear as blank cells in the coded table.

FORTRAN

The FORTRAN program for coded tables is invoked with the statement

CALL CTBL (Y, RSUB, CSUB, N, NN, NR, NC, SORTY, CHOOSE, ERR)

where

Y( )	is the N-long vector of data values;
RSUB( ), CSUB( )	are N-long integer arrays of row and column subscripts;
N	is the number of data values and, hence, the length of RSUB and CSUB as well;
NN	is the length of SORTY( )—not less than the larger of N and NR*NC;
NR	is the number of rows in the table—the integers in RSUB( ) thus count from 1 to NR;
NC	is the number of columns in the table—the integers in CSUB( ) thus count from 1 to NC;
SORTY( )	is an NN-long work array for sorting the values in Y( );
CHOOSE	is an integer flag to indicate selection when there are multiple values in a cell: 1 choose most extreme value in cell, 2 choose maximum value, or 3 choose minimum value;
ERR	is the error flag, whose values are 0 normal

71   the table has a zero dimension (NR =
        0 or NC = 0)
72   too many columns—will not fit on
        page at current margins
73   insufficient room in SORTY( ).

The FORTRAN program constructs each row of the coded table, using PUTCHR to put symbols in the output line. Each line is printed as it is completed.

BASIC

The BASIC subroutine for coded tables is entered with the N data values in Y( ), row subscripts in R( ), and column subscripts in C( ). The data are first sorted into the work array, W( ), and hinges and fences are determined. The hinges are placed in L2 and L3, the inner fences in F1 and F2, and the step (= 1.5 × H-spr) in S1. The table is then copied in row-major form into W( ). The coded table is printed cell-by-cell as it is generated.

7.5   Details and Alternatives

One obvious enhancement of a coded table is the use of color. Values above the median might be given green codes, while values below the median might have red codes. Users with more sophisticated graphics devices might prefer another choice of codes. However, it is doubtful that increasing the number of code alternatives would improve the coded table very much. Seven alternatives seems to be a comfortable number for the human mind to work with. See, for example, Miller (1956).

When the rows and columns have not only an order but also a natural or estimated spacing, it can be useful to lay out the rows and columns of the coded table according to that spacing. This is difficult to do well on a printer, but is easily accomplished with more sophisticated graphics equipment. One source of such a spacing is the row and column effects found by a median polish of the table (see Chapter 8).

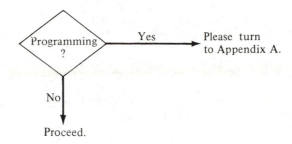

References

Berkson, J. 1958. "Smoking and Lung Cancer: Some Observations on Two Recent Reports." *Journal of the American Statistical Association* 53:28–38.

Box, G.E.P., and D.R. Cox. 1964. "An Analysis of Transformations." *Journal of the Royal Statistical Society, Series B* 26:211–243.

Miller, G.A. 1956. "The Magical Number Seven, Plus or Minus Two: Some Limits on Our Capacity for Processing Information." *Psychological Review* 63:81–97.

```
5000 REM CODED TABLE ROUTINE
5010 REM PRINTS A 7-SYMBOL CODED TABLE OF THE MATRIX IN Y()
5020 REM WITH SUBSCRIPTS IN R() AND C().
5030 REM IF THERE IS MORE THAN ONE VALUE IN ANY CELL OF THE TABLE,
5040 REM THE VALUE OF  V1  DETERMINES WHICH SHALL DETERMINE THE CODE:
5050 REM    V1=1 : THE LEAST VALUE IS CODED
5060 REM    V1=2 : THE MOST EXTREME (GREATEST MAGNITUDE) VALUE,
5070 REM    V1=3 : THE GREATEST VALUE.
5080 REM IN ALL CASES THE ENTIRE DATA SET IS USED TO FIND HINGES AND
                                                            FENCES
5090 REM THE V1=2 VERSION IS THE USUAL DEFAULT, AND IS USED IF
5100 REM  V1<>1 AND V1 <> 3.
5110 REM
5120 REM SORT Y INTO W AND GET INFORMATION ABOUT IT

5130 GOSUB 3300
5140 GOSUB 2500

5150 REM LOCAL MISSING VALUE IS ONE GREATER THAN MAX VALUE IN Y()

5160 LET E1 = W(N) + 1
5170 FOR K = 1 TO N
5180    LET W(K) = E1
5190 NEXT K

5200 REM COPY Y() TO W() INTO ROW-MAJOR FORM
5210 REM CHOOSING FROM MULTIPLE VALUES IN A CELL ACCORDING TO V1

5220 FOR K = 1 TO N
5230    LET L = C9 * (R(K) - 1) + C(K)
5240    IF W(L) = E1 THEN 5350
5250    LET W1 = W(L)
5260    LET Y1 = Y(K)
5270    IF V1 <> 1 THEN 5300
5280    IF W1 <= Y1 THEN 5360
5290    GO TO 5350
5300    IF V1 <> 3 THEN 5330
5310    IF W1 >= Y1 THEN 5360
5320    GO TO 5350

5330    REM  MOST EXTREME IS DEFAULT FOR ANY OTHER V1

5340    IF ABS(W1) >= ABS(Y1) THEN 5360
5350    LET W(L) = Y(K)
5360 NEXT K
5370 LET K = 0
```

```
5380 REM   CHARACTER SET FOR CODED TABLES IS #+.-=

5390 FOR I = 1 TO R9
5400    PRINT TAB(M0);

5410    FOR J = 1 TO C9
5420       LET K = K + 1
5430       LET X1 = W(K)
5440       IF X1 <> E1 THEN 5470
5450       PRINT " ";
5460       GO TO 5660
5470       IF X1 > L3 THEN 5590
5480       IF X1 < L2 THEN 5510
5490       PRINT ".";
5500       GO TO 5660
5510       IF X1 < F1 THEN 5540
5520       PRINT "-";
5530       GO TO 5660
5540       IF X1 < L2 - 2 * S1 THEN 5570
5550       PRINT "=";
5560       GO TO 5660
5570       PRINT "M";
5580       GO TO 5660
5590       IF X1 > F2 THEN 5620
5600       PRINT "+";
5610       GO TO 5660
5620       IF X1 > L3 + 2 * S1 THEN 5650
5630       PRINT "#";
5640       GO TO 5660
5650       PRINT "P";
5660       PRINT " ";
5670    NEXT J
5680    PRINT
5690 NEXT I
5700 PRINT
5710 RETURN
```

```
      SUBROUTINE CTBL(Y, RSUB, CSUB, N, NN, NR, NC, SORTY, CHOOSE, ERR)
C
      INTEGER N, NN, NR, NC, CHOOSE, ERR
      INTEGER RSUB(N), CSUB(N)
      REAL Y(N), SORTY(NN)
C
C  PRINT A CODED TABLE OF THE MATRIX IN Y() WITH SUBSCRIPTS IN RSUB()
C  AND CSUB().  THIS FORM OF STORING A MATRIX ALLOWS MULTIPLE DATA
C  ITEMS IN A CELL.  WHEN THERE ARE MULTIPLE DATA ITEMS IN A CELL, THIS
C  ROUTINE CONSULTS CHOOSE.  IF CHOOSE = 1, THE MOST EXTREME VALUE WILL
C  BE USED.  IF  CHOOSE = 2, THE MAXIMUM VALUE WILL BE USED.  IF
C  CHOOSE = 3, THE MINIMUM VALUE WILL BE USED.  THE FIRST CHOICE IS
C  USUALLY BEST FOR RESIDUALS.  THE SECOND AND THIRD TOGETHER CAN BE
C  VALUABLE FOR RAW DATA.
C   SORTY() MUST BE DIMENSIONED BIG ENOUGH TO CONTAIN AN ELEMENT FOR
C  EVERY CELL OF THE TABLE INCLUDING EMPTY CELLS.  THUS NN IS .GE. N.
C*** THE DIMENSIONING OF SORTY() THIS WAY DIFFERS FROM THE DESCRIPTION
C*** IN CHAPTER 7 OF ABCS OF EDA (FIRST PRINTING).
C
C  COMMON BLOCKS
C
      COMMON /CHRBUF/ P, PMAX, PMIN, OUTPTR, MAXPTR, OUNIT
      INTEGER P(130), PMAX, PMIN, OUTPTR, MAXPTR, OUNIT
C
C  LOCAL VARIABLES
      INTEGER I, J, K, IADJL, IADJH, NBIG
      INTEGER CHPT, CHMIN, CHEQ, CHM, CHPLUS, CHX, CHP, CHBL
      REAL MED, HL, HH, ADJL, ADJH, STEP, OFENCL, IFENCL, OFENCH
      REAL IFENCH, EMPTY
C
      DATA CHPT, CHMIN, CHEQ, CHM/ 46, 40, 38, 13/
      DATA CHPLUS, CHX, CHP, CHBL/ 39, 24, 16, 37/
C
C  CHECK FOR ROOM ON THE PAGE AND IN SORTY()
C
      IF(PMAX .GE. PMIN + 2*NC) GO TO 5
      ERR = 72
      GO TO 999
    5 NBIG = MAXO(N, NR*NC)
      IF(NBIG .LE. NN) GO TO 8
      ERR = 73
C *** SORTY() DIMENSIONED TOO SMALL.  THIS IS A NEW ERROR CODE.
      GO TO 999
C
C  GET SUMMARY INFORMATION ABOUT DATA IN TABLE
C
    8 IF(NR .GT. 0  .AND.  NC .GT. 0) GO TO 10
      ERR = 71
      GO TO 999
   10 DO 20 K = 1, N
         SORTY(K) = Y(K)
   20 CONTINUE
      CALL YINFO(SORTY, N, MED, HL, HH, ADJL, ADJH, IADJL, IADJH, STEP,
     1 ERR)
```

```
          IF (ERR .NE. 0) GO TO 999
          OFENCL = HL - 2.0*STEP
          IFENCL = HL - STEP
          OFENCH = HH + 2.0*STEP
          IFENCH = HH + STEP
C
C   SET INTERNAL EMPTY CODE GREATER THAN THE  LARGEST VALUE
C   TO BE SURE IT IS UNIQUE.  IF IT IS NEGATIVE, SET EMPTY POSITIVE.
          EMPTY = ABS(SORTY(N)) * 1.1 + 1.0
C   WE NO LONGER NEED THE SORTED VERSION, SO RE-USE THE SPACE IN SORTY()
          DO 22 K = 1, NBIG
             SORTY(K) = EMPTY
       22 CONTINUE
C
C   TRANSFER DATA FROM Y() INTO SORTY() IN ROW MAJOR FORMAT -- THAT IS,
C   ITEMS IN THE FIRST ROW FROM LEFT TO RIGHT, FOLLOWED BY THE SECOND
C   ROW (LEFT TO RIGHT), AND SO ON.  IF TWO DATA ITEMS ARE FOUND IN THE
C   SAME CELL, KEEP THE ONE INDICATED BY  CHOOSE.
C
          DO 30 K = 1, N
             I = NC * (RSUB(K) -1) + CSUB(K)
             IF(SORTY(I) .EQ. EMPTY) GO TO 25
             IF(CHOOSE .EQ. 1  .AND.  ABS(SORTY(I)) .GE. ABS(Y(K))) GO TO 30
             IF(CHOOSE .EQ. 2  .AND.  SORTY(I) .GE. Y(K)) GO TO 30
             IF(CHOOSE .EQ. 3  .AND.  SORTY(I) .LE. Y(K)) GO TO 30
       25    SORTY(I) = Y(K)
       30 CONTINUE
          K = 0
          DO 50 I = 1, NR
             DO 40 J = 1, NC
                K = K+1
                IF(SORTY(K) .EQ. EMPTY) CALL PUTCHR(0, CHBL, ERR)
                IF(SORTY(K) .EQ. EMPTY) GO TO 35
                IF(SORTY(K) .LT. OFENCL) CALL PUTCHR(0, CHM, ERR)
                IF((SORTY(K) .GE. OFENCL) .AND. (SORTY(K) .LT. IFENCL))
        1        CALL PUTCHR(0, CHEQ, ERR)
                IF((SORTY(K) .GE. IFENCL) .AND. (SORTY(K) .LT. HL))
        1        CALL PUTCHR(0, CHMIN, ERR)
                IF((SORTY(K) .GE. HL) .AND. (SORTY(K) .LE. HH))
        1        CALL PUTCHR(0, CHPT, ERR)
                IF((SORTY(K) .GT. HH) .AND. (SORTY(K) .LE. IFENCH))
        1        CALL PUTCHR(0, CHPLUS, ERR)
                IF((SORTY(K) .GT. IFENCH) .AND. (SORTY(K) .LE. OFENCH))
        1        CALL PUTCHR(0, CHX, ERR)
                IF(SORTY(K) .GT. OFENCH) CALL PUTCHR(0, CHP, ERR)
       35       CALL PUTCHR(0, CHBL, ERR)
                IF(ERR .NE. 0) GO TO 999
       40    CONTINUE
             CALL PRINT
       50 CONTINUE
          CALL PRINT
      999 RETURN
          END
```

Chapter 8

Median Polish

The coding technique of Chapter 7 displays two-way tables of data and reveals patterns in these tables. Such graphical displays are important, but often they invite us to analyze the data—to summarize the overall pattern simply and examine the residuals it leaves behind. To summarize a pattern in a table, we must find a way to characterize the patterns that we are likely to encounter in two-way tables. Median polish is a simple method for discovering a common type of pattern.

8.1   Two-Way Tables

Patterns in two-way tables are often described in terms of differences among entire rows or columns of data values. Thus, a row with larger-than-average data values might be noted. We often label each data value in a cell of a two-way table with the number of the row and the number of the column in which the value appears, and we think of the row and column identities as *factors* that help us to account for observed patterns. For example, the data

factors

219

value in the second row and third column of a table is denoted by $y_{2,3}$. More generally, the data value in the *i*th row and *j*th column is denoted by $y_{i,j}$.

response

While the rows and columns are the factors helping to describe the data values in the table, the data values themselves are thought of as the *response*. This dichotomy is much the same as we saw for fitting lines in Chapter 5—where x was the factor and y was the response—and for smoothing in Chapter 6—where t was the factor and y was the response. In each model we attempt to describe the response, y, using the factors, and we know that the description cannot be expected to fit the observed data exactly.

The death rate table we examined in Chapter 7 provides a convenient example. The data shown in Exhibit 7–1 are repeated in Exhibit 8–1. Here the response is the death rate, and the two factors are the cause of death and the average amount of tobacco smoked. As the data are laid out, the rows correspond to the causes, and the columns correspond to the extent of smoking. We naturally expect some causes to be responsible for many more deaths than others. The row medians in Exhibit 8–1 and the coded table in Exhibit 7–2 both reveal higher death rates from coronary thrombosis, other cardiovascular diseases, and cerebral hemorrhage, and lower rates from upper respiratory cancer, pulmonary TB, and peptic ulcer. In light of today's knowledge, we would expect smoking to affect the death rate for several causes. If this pattern is present, it is not obvious, but we may be able to judge more clearly after adjusting for the differences among the typical death rates for the various causes.

8.2   A Model for Two-Way Tables

When we chose a model to describe x-y data in Chapter 5, we used a straight line because of its simplicity. Two-way tables require a different kind of model because they involve three components—the row factor, the column factor, and the response—but we still aim for simplicity.

The straight line is a convenient model for y versus x because it fits each y-value with the sum of two simple components: a constant intercept value to anchor the line where $x = 0$ and a slope multiplied by x to account for changes in y associated with changes in x away from $x = 0$. Because these two components are added in the fit, we can polish the resistant line by adding adjustments to the slope and intercept.

Exhibit 8–1   Male Death Rates per 1000 by Cause of Death and Average Amount of Tobacco Smoked Daily

	Amount of Tobacco Smoked				
Cause of Death	*None*	*1–14 Grams*	*15–24 Grams*	*25+ Grams*	*Row Median*
Cancers					
Lung	0.07	0.47	0.86	1.66	.665
Upper respiratory	0.00	0.13	0.09	0.21	.11
Stomach	0.41	0.36	0.10	0.31	.335
Colon and rectum	0.44	0.54	0.37	0.74	.49
Prostate	0.55	0.26	0.22	0.34	.30
Other	0.64	0.72	0.76	1.02	.74
Respiratory diseases					
Pulmonary TB	0.00	0.16	0.18	0.29	.17
Chronic bronchitis	0.12	0.29	0.39	0.72	.34
Other	0.69	0.55	0.54	0.40	.545
Coronary thrombosis	4.22	4.64	4.60	5.99	4.62
Other cardiovascular	2.23	2.15	2.47	2.25	2.24
Cerebral hemorrhage	2.01	1.94	1.86	2.33	1.975
Peptic ulcer	0.00	0.14	0.16	0.22	.15
Violence	0.42	0.82	0.45	0.90	.635
Other diseases	1.45	1.81	1.47	1.57	1.52

Note: Rates are not age-adjusted.

additive model
common value
row effects
column effects

For two-way tables we use a similar *additive model,* which represents each cell of the table as the sum of three simple components: a constant *common value* to summarize the general level of y, *row effects* to account for changes in y from row to row relative to the common value, and *column effects* to account for changes in y from column to column relative to the common value. Exhibit 8–2 shows an example that displays the three components of an additive fit. As shown there, each component describes a table with very simple structure—constant, or with constant stripes across rows, or with constant stripes down columns.

The common term, 8 in Exhibit 8–2, describes the level of the data values in the table as a whole. It can thus be thought of as describing a two-way table that has the same constant value in each cell. Each row effect

Exhibit 8–2   The Components of an Additive Model for a Two-Way Table

(a) Common Term

8	8	8	
8	8	8	
8	8	8	
8	8	8	
8	8	8	
			8

The common term fits a constant for each cell of the table—in this case 8.

(b) Row Effects

6	6	6	6
−1	−1	−1	−1
0	0	0	0
4	4	4	4
−8	−8	−8	−8

The row effects fit the difference between each row and the common term. They fit a table of adjustments that is constant across each row.

(c) Column Effects

0	−3	0
0	−3	0
0	−3	0
0	−3	0
0	−3	0
0	−3	0

The column effects fit the difference between each column and the common term. They fit a table of adjustments that is constant down each column.

(d) Sum

14	11	14	6
7	4	7	−1
8	5	8	0
12	9	12	4
0	−3	0	−8
0	−3	0	8

The full fit is the sum of tables a, b, and c above. The value in row *i* and column *j* is found from fit_{ij} = common + row_i + col_j. Example: $\hat{y}_{1,2} = 11 = 8 + 6 + (-3)$.

describes the way in which the data values in its row tend to differ from the common level. The collection of row effects thus describes a table that is constant across each row. Similarly, the column effects describe the way in which the data values in each column tend to differ from the common level. They thus describe a table that is constant down each column. The sum of these three components—common term, row effects, and column effects—can be found by adding the three simple tables together. Each cell of this summed table describes, or fits, the corresponding cell of the original table of data. Thus the fit for the cell in row *i* and column *j* is

$$\text{fit}_{ij} = \text{common term} + \text{row effect}_i + \text{column effect}_j.$$

An additive fit to an *R*-row and *C*-column table uses 1 common value, *R* row effects, and *C* column effects to describe $R \times C$ data values. More important than the use of fewer numbers, each of the components is likely to show understandable regularities.

The additive model provides a precise way of describing the patterns that we look for in a coded table. For example, if the columns have a natural order and the coding shows a trend across the columns, then the column effects will describe this trend in numerical terms. If the rows have no natural order, we may still want to examine the differences among them; and the row effects would form the basis for this examination.

8.3   Residuals

Whenever we fit a model to data, we need to examine the differences between the raw data and the values suggested by the fitted equation. For additive models fitted to two-way tables, we can find these differences from

$$\text{residual}_{ij} = \text{data}_{ij} - \text{fit}_{ij}$$

or, equivalently,

$$\text{residual}_{ij} = \text{data}_{ij} - (\text{common} + \text{row effect}_i + \text{column effect}_j).$$

We can rearrange the equation as

$$\text{data}_{ij} = \text{common} + \text{row effect}_i + \text{column effect}_j + \text{residual}_{ij}.$$

There is a residual for each original data value, so the residuals themselves are a table having the same number of rows and the same number of columns as the original data table.

Exhibit 8–3 shows a two-way table of deaths from sport parachuting in each of three years according to the experience of the parachutist. The additive model displayed in Exhibit 8–2 is, in fact, an additive fit for these data. Exhibit 8–3c shows the residuals as the final component of the description of the data. The three components in Exhibit 8–2 form the fit, and the table of residuals shows how well this fit describes the data. We see, for

Exhibit 8–3   Deaths from Sport Parachuting

(a) The Data

	Year		
Number of Jumps	*1973*	*1974*	*1975*
1–24	14	15	14
25–74	7	4	7
75–199	8	2	10
200 or more	15	9	10
unreported	0	2	0

(b) The Fit (from Exhibit 8–2d)

14	11	14	6
7	4	7	−1
8	5	8	0
12	9	12	4
0	−3	0	−8
0	−3	0	8

(c) The Residuals

0	4	0	6
0	0	0	−1
0	−3	2	0
3	0	−2	4
0	5	0	−8
0	−3	0	8

Source: Data from Metropolitan Life Insurance Company, *Statistical Bulletin* 60, no. 3 (1979). p. 4. Reprinted by permission.

Note: $data_{ij}$ = common + row_i + col_j + $resid_{ij}$. Example: $y_{1,2} = 15 = 8 + 6 + (−3) + 4$

example, that the fitted value of 11 deaths for inexperienced parachutists in 1974 was too low by 4—actually there were 15 fatalities in that category that year.

The residuals from an additive fit often reveal patterns that are not readily apparent in the original data. A row or column that fails to follow a general pattern established by other rows or columns will produce a prominent residual pattern. A single extraordinary value in the table will, when we fit the model by median polish, leave a large residual. It is usually worthwhile to examine a coded table of the residuals to look for patterns.

8.4   Fitting an Additive Model by Median Polish

There are many ways to find an additive model for a two-way table. Regardless of the method, we must progress from the original data table to (1) a common value, (2) a set of row effects, (3) a set of column effects, and (4) a table of residuals, all of which sum to the original data values. Several methods do this in stages, sweeping information on additive behavior out of the data and into the common term, row effects, and column effects in turn. If each stage ensures that the sum of the fit components and the residuals equals the original data, then the result of several stages will also be additive.

In Chapters 5 and 6 we protected our fits from the effects of extraordinary data values by summarizing appropriate portions of the data with medians. We can do the same for two-way tables, using medians in each stage of the fitting process to summarize either rows or columns, and sweeping the information they describe into the fit.

For example, we can begin by finding the median of the numbers in a row of the table, subtracting it from all the numbers in that row, and using it as a partial description for that row. This operation sweeps a contribution from the row into the fit. We do this for each row, producing a column of row medians and a new table from which the row medians have been subtracted. (Consequently, the median of each row in this new table is zero.) The operation just described is portrayed in Exhibit 8–4a, where the first box represents the data, and the arrows across the box indicate the calculation of row medians. The subtraction of these row medians from the data values completes Sweep 1 (Exhibit 8–4b). At this stage, the column of original row medians serves as a partial row description and occupies the position of the row effects—to the right of the main box.

Row medians for the death rate data were shown in Exhibit 8–1. The results of Sweep 1 on the same data are shown in Exhibit 8–5, which repeats the original column of row medians. We saw, in Exhibit 8–1, that, for example, the death rate from stomach cancer among men who smoked an average of 1–14 grams of tobacco per day ($y_{3,2}$) was 0.36. The median death rate from stomach cancer across all four columns is .335. The residual in

Exhibit 8–4   Median Polish as a Sequence of Four Sweeping Operations, Starting with the Rows
of the Data

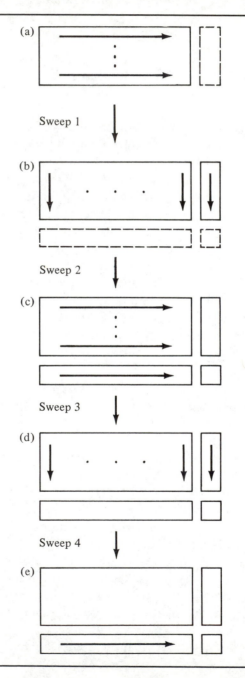

Exhibit 8–5 Result of Sweep 1, Removing Row Medians throughout Exhibit 8–1, also Showing Column Medians

	0	1–14	15–24	25+	Part
	−.595	−.195	.195	.995	.665
	−.11	.02	−.02	.10	.11
	.075	.025	−.235	−.025	.335
	−.05	.05	−.12	.25	.49
	.25	−.04	−.08	.04	.30
	−.10	−.02	.02	.28	.74
	−.17	−.01	.01	.12	.17
	−.22	−.05	.05	.38	.34
	.145	.005	−.005	−.145	.545
	−.40	.02	−.02	1.37	4.62
	−.01	−.09	.23	.01	2.24
	.035	−.035	−.115	.355	1.975
	−.15	−.01	.01	.07	.15
	−.215	.185	−.185	.265	.635
	−.07	.29	−.05	.05	1.52
Median	−.10	−.01	−.02	.12	.545

Exhibit 8–5, .025, is found as

$$0.36 - .335 = .025.$$

The column of row medians is labeled "Part" in Exhibit 8–5 because of its role as a partial description. In preparation for the next operation, Exhibit 8–5 also records the median of the numbers now in each column, as well as the median of the column of row medians.

We turn next to the columns, acting now on the table of residuals left by the first sweep. We find the median of each column (already recorded in Exhibit 8–5). Then we subtract each column median from the numbers in its column and use it as the partial description for that column. In addition, we find the median of the column of row descriptions, subtract it from each row description, and use this median as a partial common value. These steps constitute Sweep 2 in the schematic diagram. Note that the rectangles bordering the third main box in Exhibit 8–4 include two new parts, which occupy the positions of the column effects and the common value. For the death rate data, Exhibit 8–6 shows the result of Sweep 2.

Continuing with the value in row 3, column 2, we now have the column-2 effect of $-.01$ and a common term of .545, and the row-3 effect of .335 has had the common term subtracted from it, yielding $-.21$. Removing all of these components leaves a new residual of .035. The data value $y_{3,2}$ is then summarized at this step as

$$y_{3,2} = \text{common} + \text{row effect}_3 + \text{column effect}_2 + \text{residual}_{3,2}$$

or

$$0.36 = \quad .545 \quad - \quad .21 \quad - \quad .01 \quad + \quad .035.$$

We prepare for the next step by recording the median of each row in Exhibit 8–6, including the row of partial column descriptions, at the right of the table in the column headed "Median." The $-.015$ at the intersection of the "Part" row and the "Median" column is the median of the row of partial column descriptions; it will be used to adjust the common term.

Exhibit 8–6 Result of Sweep 2, Removing Column Medians throughout Exhibit 8–5, also Showing Row Medians

	0	*1–14*	*15–24*	*25+*	*Median*	*Part*
	−.495	−.185	.215	.875	.015	.12
	−.01	.03	0	−.02	−.005	−.435
	.175	.035	−.215	−.145	−.055	−.21
	.05	.06	−.10	.13	.055	−.055
	.35	−.03	−.06	−.08	−.045	−.245
	0	−.01	.04	.16	.02	.195
	−.07	0	.03	0	0	−.375
	−.12	−.04	.07	.26	.015	−.205
	.245	.015	.015	−.265	.015	0
	−.30	.03	0	1.25	.015	4.075
	.09	−.08	.25	−.11	.005	1.695
	.135	−.025	−.095	.235	.055	1.43
	−.05	0	.03	−.05	−.025	−.395
	−.115	.195	−.165	.145	.015	.09
	.03	.30	−.03	−.07	0	.975
Part	−.10	−.01	−.02	.12	−.015	.545

At this stage, after a sweep across the rows and a sweep down the columns, we could stop; but, because we are using the median, it will usually be possible to improve the partial descriptions of the data by performing another sweep across the rows and another sweep down the columns. Earlier, just after Sweep 1, each row in the remaining table of numbers had a median of zero. However, this may not be true after Sweep 2; so sweeping the rows of the table of residuals left after Sweep 2 yields some adjustments that will improve the partial row descriptions and reduce the overall size of the residuals. (Of course, not every residual will be made smaller. Some may grow substantially. But overall, most residuals will be brought closer to zero by performing additional sweeps.)

Sweep 3 repeats Sweep 1, except that the row medians found are added to the previous row descriptions. Sweep 3 also finds the median of the column descriptions, subtracts this median from each column description, and adds it to the common value. Exhibit 8–7 demonstrates Sweep 3 for the death rate data.

Exhibit 8–7 Result of Sweep 3, Removing Row Medians throughout Exhibit 8–6, also Showing Column Medians

	0	*1–14*	*15–24*	*25+*	*Part*
	−.51	−.20	.20	.86	.135
	−.005	.035	.005	−.015	−.44
	.23	.09	−.16	−.09	−.265
	−.005	.005	−.155	.075	0
	.395	.015	−.015	−.035	−.29
	−.02	−.03	.02	.14	.215
	−.07	0	.03	0	−.375
	−.135	−.055	.055	.245	−.19
	.23	0	0	−.28	.015
	−.315	.015	−.015	1.235	4.09
	.085	−.085	.245	−.115	1.70
	.08	−.08	−.15	.18	1.485
	−.025	.025	.055	−.025	−.42
	−.13	.18	−.18	.13	.105
	.03	.30	−.03	−.07	.975
Median	−.005	.005	0	0	.015
Part	−.085	.005	−.005	.135	.53

Now, for example, the median of the numbers remaining in row 3 is −.055. This median is subtracted from each number in row 3 and added to the row effect (−.21) to obtain a new row effect, −.265. The median of the row of column medians, −.015, has also been subtracted from each column median and added to the common term. The new description for $y_{3,2}$ is

$$0.36 = 0.53 − .265 + .005 + .09.$$

(We note that although the residual in this cell is actually growing at each step, the residuals in the table are generally getting smaller.)

Sweep 4 parallels Sweep 2, working again with the columns instead of the rows. Exhibit 8–8 shows the result for the death rate data. This takes us to the bottom in the schematic view of the process in Exhibit 8–4.

Only one detail remains: We find the median of the adjusted column

Exhibit 8–8 Result of Sweep 4, Removing Column Medians throughout Exhibit 8–7 (completing the standard median polish for these data)

	None	1–14	15–24	25+	Effect
Cancers					
Lung	−.505	−.205	.20	.86	.12
Upper respiratory	0	.03	.005	−.015	−.455
Stomach	.235	.085	−.16	−.09	−.28
Colon and rectum	0	0	−.155	.075	−.015
Prostate	.40	.01	−.015	−.035	−.305
Other	−.015	−.035	.02	.14	.20
Respiratory diseases					
Pulmonary TB	−.065	−.005	.03	0	−.39
Chronic bronchitis	−.13	−.06	.055	.245	−.205
Other	.235	−.005	0	−.28	0
Coronary thrombosis	−.31	.01	−.015	1.235	4.075
Other cardiovascular	.09	−.09	.245	−.115	1.685
Cerebral hemorrhage	.085	−.085	−.15	.18	1.47
Peptic ulcer	−.02	.02	.055	−.025	−.435
Violence	−.125	.175	−.18	.13	.09
Other diseases	.035	.295	−.03	−.07	.96
Effect	−.09	.01	−.005	.135	.545

Note: In this example, the median of the (adjusted) partial column descriptions is zero (to working accuracy), so they become the column effects.

descriptions and add it to the common value. (In Exhibit 8–8, this adjustment turns out to have no effect because, to the 2-decimal-place accuracy of the data, the median of the column descriptions is zero.) This step ensures that the column effects will have a median of zero. (The row effects were left with a zero median by Sweep 4.) We could instead continue to sweep the rows and the columns alternately, looking for further adjustments, but such adjustments are generally much smaller than the ones found in Sweep 3 and Sweep 4, and sometimes they are exactly zero and thus would not change the fit. Therefore, the standard version of median polish stops after Sweep 4. The fit for some tables may improve sufficiently with additional steps to make them worthwhile. Especially when we have a computer to do the work, we may choose to try a few extra steps. One sweep across the rows or the columns is also known *half-step* as a *half-step*; and a pair of sweeps, working with both the rows and the *full-step* columns, constitutes a *full-step*.

Because we have swept the common term out of the partial row and column descriptions at each stage, what we have left are adjustments relative to the common term. They are thus the row and column effects we need for the additive model.

For the death rate data, the calculations have brought us to the point where, in Exhibit 8–8, we need only affix the label "effect" to the partial descriptions for the rows and the columns. The numbers left in the table, where the data values were originally, are the residuals. The pieces of the additive fit are arranged around the edge of that table: an effect for each row, an effect for each column, and the common value. Thus, the fitted death rate from stomach cancer among men who smoked an average of 1–14 grams of tobacco per day (the $y_{3,2}$ value) is

$$.545 + (-.28) + .01 = .275,$$

and the residual is

$$.36 - .275 = .085.$$

We can easily check that in each cell of Exhibit 8–8 the fitted value and the residual add up to the data value.

Now that we have the pieces of the fit, what do they tell us? The common value is .545 deaths per 1000 men. This is not a death rate for the population, but rather a typical death rate for these causes among men with this range of smoking habits. The common value serves us primarily as a standard against which to measure patterns.

The effect values for cause of death lead us to qualify our earlier impression of substantial variation. Coronary thrombosis (at 4.075 deaths/

1000 above the common level) is clearly a major killer. However, except for the cardiovascular diseases, most causes show effects close to the common level. (The largest remaining effect is for "Other diseases," which is clearly a catchall and not a specific cause of death.) The effects for amount of smoking are smaller and range only from −.09 to .135. It seems from these effects that heavy smokers, 25+ grams per day, are somewhat more at risk than non-smokers. We do not, however, expect smoking to have the same impact on death rates for all causes. Indeed, we would be surprised if smoking had much to do with death by violence. If the effect of smoking on the death rate from a particular cause does not conform to the overall pattern in the column effects, this fact would have to show up in the residuals for its row of the table.

We usually look at the residuals to find such remaining patterns or any unusual values, and we often construct a coded table such as Exhibit 8–9, which displays the residuals from Exhibit 8–8. The strongest pattern is that of lung cancer, which shows a steadily rising death rate with increased smoking. This pattern indicates that the impact of smoking on death rates from lung cancer is much stronger than the slight overall increase we observed in the column effects. Even after allowing for higher death rates among smokers across all causes, lung cancer death rates show a greater change—non-smokers die from lung cancer less frequently than we might otherwise predict, and heavy smokers die from lung cancer much more often.

The pattern for coronary thrombosis is similar, if less consistent. However, here the coding in Exhibit 8–9 has partially hidden a truly extraordinary residual. The residual for the death rate of heavy smokers from coronary thrombosis is a remarkable 1.235 deaths per 1000 men—larger than any of the *death rates* from specific non-cardiovascular diseases. That is, the death rate from coronary thrombosis is increased by heavy smoking over the (already large) value we would predict for this cause of death (even after allowing for generally higher death rates observed for heavy smokers), and the amount of the increase is greater than the death rate from most diseases.

The other noteworthy positive residual is the residual for deaths from prostate cancer among non-smokers. It might appear that we have discovered a hazard of *not* smoking, but another explanation seems more likely. Prostate cancer is a disease generally afflicting older men. It is likely that, before they reach the age at which prostate cancer is common, a larger number of smokers have already succumbed to other diseases. Thus fewer smokers than non-smokers remain to face the risk of dying from prostate cancer.

One major reason for using medians in finding the additive fit was to protect our results from being distorted by extraordinary values. Although some of the examples in earlier chapters have included extreme values that seemed wrong or out of place, the data values in the death rates example are

Exhibit 8–9   The Residuals (from Exhibit 8–8) of the Median Polish of Death Rates, Coded and Displayed

	None	*1–14*	*15–24*	*25+*	*Effect*
Cancers					
Lung	=	−	+	P	.12
Upper respiratory	·	·	·	·	−.455
Stomach	+	·	−	−	−.28
Colon and rectum	·	·	−	·	−.015
Prostate	#	·	·	·	−.305
Other	·	·	·	+	.20
Respiratory diseases					
Pulmonary TB	·	·	·	·	−.39
Chronic bronchitis	−	·	·	+	−.205
Other	+	·	·	−	0
Coronary thrombosis	=	·	·	P	4.075
Other cardiovascular	+	−	+	−	1.685
Cerebral hemorrhage	·	−	−	+	1.47
Peptic ulcer	·	·	·	·	−.435
Violence	−	+	−	+	.09
Other diseases	·	+	·	−	.96
Effect	−.09	.01	−.005	.135	.545

more or less what we might expect, and the resistance of the median has allowed the three large residuals to become prominent.

One last comment on median polish: We could have chosen to begin median polish with columns instead of rows. The procedure is essentially the same, but the resulting fit may be slightly different. For purposes of exploration, the difference does not matter. When we can use a computer to do the work, we may want to try both forms and compare the results.

8.5 Re-expressing for Additivity

Often a table that is not described well by an additive model can be made more nearly additive by re-expressing the data values. When we used re-expression to straighten a bend in y versus x, it was easy to see the bend in a plot. In a table, the simplest kind of "bending" that cannot be described by an additive model is a twisting of the corners: one diagonally opposite pair of corners too high and the other diagonally opposite corners too low, when the rows and columns are in order according to their effects in an additive fit. We can return to Exhibit 8–2 to see why such a pattern cannot be fit by an additive model. If, for example, the two corners at the top of the table were high, the effects for the top rows could be increased to make the additive model fit better. However, a pattern of diagonally opposite high or low values cannot be accounted for by any of the three components of the additive fit nor by any additive combination of them.

When the data values follow such a "saddle" pattern, the diagonally opposite corners of the table of residuals will have the same sign. Exhibit 8–10 shows the two possible types of saddle-shaped residual patterns. Here the signs of the effects are shown in the borders of the table and used to partition the table of residuals into four regions. The signs shown for these regions summarize the signs of the residuals. Evidence of such a pattern—for example, in a coded table of the residuals—suggests that a well-chosen re-expression is likely to help. Later in this section we consider how to make this choice simply.

Exhibit 8–11 reports the time taken by the winning runner in five

Exhibit 8–10   The Two Types of Residual Patterns that Suggest Re-expression to Promote Additivity in a Two-Way Table

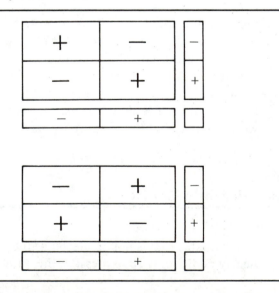

Exhibit 8–11   Winning Time in Men's Olympic Runs by Year and Distance (unit = .1 sec.)

Year	100m	200m	400m	800m	1500m
			Distance		
1948	103	211	462	1092	2298
1952	104	207	459	1092	2252
1956	105	206	467	1077	2212
1960	102	205	449	1063	2156
1964	100	203	451	1051	2181
1968	99	198	438	1043	2149
1972	101	200	447	1059	2163

Source: Data from *The World Almanac* (New York: Newspaper Enterprise Association, Inc., 1973) p. 858. Reprinted by permission.

men's track events at the Olympic Games from 1948 to 1972. The five events are the 100-, 200-, 400-, 800-, and 1500-meter runs. Although the length of the run greatly influences a runner's strategy for the race, we can begin by analyzing winning time in relation to year and distance. Exhibit 8–11 presents the data (in units of .1 second to eliminate the decimal point and make residuals easier to scan for patterns), and Exhibit 8–12 shows an analysis by median polish. When we rearrange the rows of Exhibit 8–12 to put the years in

Exhibit 8–12   Median-Polish Analysis of Winning Times of Exhibit 8–11 (unit = .1 sec.)

Year	100m	200m	400m	800m	1500m	Effect
1948	−10	−4	0	18	104	11
1952	−6	−5	0	21	61	8
1956	−11	−12	2	0	15	14
1960	0	1	−2	0	−27	0
1964	0	1	2	−10	0	−2
1968	10	7	0	−7	−21	−13
1972	3	0	0	0	−16	−4
Effect	−349	−247	0	612	1732	451

the same order as their effects, the opposite-corners sign pattern of the residuals is quite evident (Exhibit 8–13).

We can use the pieces of the additive fit to approximate the pattern of the residuals. The negative residuals are generally associated with row effects and column effects that have opposite signs, while the positive residuals are associated with row effects and column effects that have the same sign. To judge the strength of this pattern of association, we compute a *comparison value* for each cell of the table:

comparison value

$$c_{ij} = \frac{(\text{row effect}_i) \times (\text{column effect}_j)}{\text{common}}.$$

A comparison value, c_{ij}, found in this way will generally have the same sign as the corresponding residual because row and column effects with opposite signs will generate negative comparison values, while same-sign effects will generate positive comparison values. Moreover, if the saddle-shaped pattern in the residuals is more pronounced in the corners, where the effects have greater magnitude, the more extreme comparison values will correspond to the more extreme residuals.

As we saw in the death rates example, median polish can allow an occasional extraordinary residual. Consequently, a resistant line is a good choice for summarizing the relationship between residual$_{ij}$ and c_{ij}, since it will not be influenced unduly by a few extraordinary residuals. Exhibit 8–14 gives the table of comparison values corresponding to Exhibit 8–13. Exhibit 8–15 shows the plot of each residual against its comparison value. Several points in

Exhibit 8–13 Rows of Exhibit 8–12 Rearranged to Put Row Effects into Order

Year	100m	200m	400m	800m	1500m	Effect
1968	10	7	0	−7	−21	−13
1972	3	0	0	0	−16	−4
1964	0	1	2	−10	0	−2
1960	0	1	−2	0	−27	0
1952	−6	−5	0	21	61	8
1948	−10	−4	0	18	104	11
1956	−11	−12	2	0	15	14
Effect	−349	−247	0	612	1732	451

Exhibit 8–14 Comparison Values Corresponding to the Residuals in Exhibit 8–13

Year	100m	200m	400m	800m	1500m
1968	10.1	7.1	0	−17.6	−49.9
1972	3.1	2.2	0	−5.4	−15.4
1964	1.5	1.1	0	−2.7	−7.7
1960	0	0	0	0	0
1952	−6.2	−4.4	0	10.9	30.7
1948	−8.5	−6.0	0	14.9	42.2
1956	−10.8	−7.7	0	19.0	53.8

the plot stray noticeably, but a straight line with slope equal to 1 seems to be a reasonable way to start summarizing the relation between residuals and comparison values.

Because the plot suggests that, roughly,

$$residual = comparison\ value,$$

Exhibit 8–15 Plot of Residuals against Comparison Values for the Winning Times

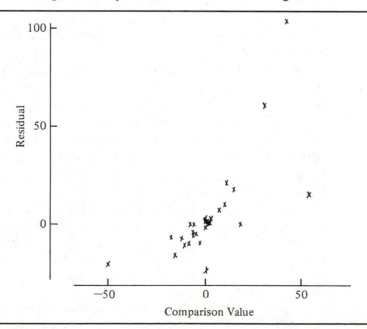

one very simple action is possible. We could add the comparison values to our additive model (and subtract them from the residuals) to get a better description of the data:

$$\text{data}_{ij} = \text{common} + \text{row effect}_i + \text{column effect}_j$$

$$+ \frac{(\text{row effect}_i) \times (\text{column effect}_j)}{\text{common}} + \text{residual}_{ij}.$$

However, we usually use the line relating residuals to comparison values as a guide for selecting a re-expression instead.

The extended model (in the previous equation) including the comparison values could be rewritten as

$$\text{data}_{ij} = \text{common} \times \left(1 + \frac{\text{row effect}_i}{\text{common}}\right) \times \left(1 + \frac{\text{column effect}_j}{\text{common}}\right) + \text{residual}_{ij}.$$

As we noted in Section 8.1, we prefer models in which the pieces add rather than multiply; so we are led to try re-expressing by logarithms because

$$\log(a \times b \times c) = \log(a) + \log(b) + \log(c).$$

Exhibit 8–16 shows the logs of the Olympic runs data, and Exhibit 8–17 shows the additive model and the residuals obtained by median polish. The analysis is clearly improved; almost all the residuals are quite small. Thus we can focus most of our attention on the fit. Because adding a constant to the logarithm of a number is equivalent to multiplying the number by a constant, the additive analysis for the log re-expression is not difficult to interpret. For example, the column effects indicate that the winning time for the 1500m run is typically about five times that for the 400m run. (Algebraically, log(1500m effect) = log(400m effect) + .690 = log(400m effect) + log(4.9); so the 1500m effect is roughly equal to the 400m effect times 4.9.) Beyond the fact that the column effects increase steadily with the length of the race, the differences between adjacent effects are almost constant. It would seem that a doubling of race length leads to slightly more than a doubling of time. (Because log(2) = .301, a constant effect difference of 301 for the first four races would have indicated a doubling of time.) To look further, we might plot the column effect against the log of the race length.

We might also plot the row effects against the year of the Olympiad.

Exhibit 8–16 Logarithm of Winning Time in Men's Olympic Runs (unit = .001)

			Distance		
Year	100m	200m	400m	800m	1500m
1948	1013	1324	1665	2038	2361
1952	1017	1316	1662	2038	2353
1956	1021	1314	1669	2032	2345
1960	1009	1312	1652	2027	2334
1964	1000	1307	1654	2022	2339
1968	996	1297	1642	2018	2332
1972	1004	1301	1650	2025	2335

Note: Original data in Exhibit 8–11.

The pattern is a reasonably steady downtrend, but we would want to look further into 1968 and 1972. (Perhaps the altitude or other conditions in Mexico City, site of the 1968 Olympic Games, were responsible for the remarkably fast races.)

The technique of plotting residuals against comparison values can guide us to re-expressions other than the log. In general, once we find the slope, b, relating the residuals to the comparison values, the quantity $(1 - b) = p$ is a good estimate of the power we should try. If the plot has zero

Exhibit 8–17 Median-Polish Analysis of Logarithm of Winning Time in Exhibit 8–16 (unit = .001)

Year	100m	200m	400m	800m	1500m	Effect
1948	−6	4	0	0	6	11
1952	0	−2	−1	2	0	9
1956	8	0	10	0	−4	5
1960	1	3	−2	0	−10	0
1964	−3	3	5	0	0	−5
1968	0	0	0	3	0	−12
1972	0	−4	0	2	−5	−4
Effect	−646	−345	0	373	690	1654

slope ($b = 0$), then $p = 1$, and no re-expression is needed. In our example, b was nearly 1; so $p = 0$, and we chose the log. (Recall from Section 2.4 that the log plays the role of the zero power in the ladder of powers.) In finding the slope, it is important to use judgment, as well as a technique such as the resistant line (Chapter 5), which will not be affected by the large residuals that median polish can leave when a data value is unusual. The combined process—median polish, the plot of residuals against comparison values, and then the resistant line—makes the search for a re-expression quite resistant to outliers.

8.6   Median Polish from the Computer

Iterative techniques such as median polish are often easier to program for a computer than to do by hand. The programs at the end of this chapter require that the data table be specified in three parallel arrays: one array for data, one for row numbers, and one for column numbers. (For a detailed description of this format, see Section 7.3.) These programs compute the row effects, column effects, common term, and residuals, but they do not print out any of these results. The best methods for displaying the results as tables depend upon the computer system being used; any simple programs provided here would have had difficulty with large tables. Nevertheless, the array of residuals returned by the programs is in an appropriate form for the coded-table programs discussed in Chapter 7.

When we use the computer, we can consider analyzing more complex tables. For example, the programs allow for empty cells in a table. The effect for any row or column containing an empty cell is based on the remaining non-empty cells. A fitted value can be found for an empty cell, but no residual can be computed.

Although median polish is an iterative procedure, no convergence check to stop the iteration automatically is included. Instead, users of the programs must specify the number of sweeps or half-steps. For data exploration, four half-steps seems adequate in most situations.

In addition, users must choose whether to remove medians from rows or columns first. For some data, the final fit and residuals will differ when these two starts are compared. Although it is quite rare for the gross structure of the fitted models to differ in important ways, the availability of machine-computed median polish makes it practical to find both versions and compare them.

* 8.7   Median Polish and ANOVA

Readers who are acquainted with the two-way analysis of variance (ANOVA) will have noticed that median polish and two-way ANOVA both start with the same data. The two-way ANOVA uses the same additive model as median polish, but it fits this model by finding row and column *means*. The difference between median polish and ANOVA is related to the difference between the resistant line and least-squares regression (Section 5.10). The exploratory techniques are resistant to outliers and require iterative calculations. However, they do not as yet provide any hypothesis-testing mechanisms.

Statistically sophisticated readers may wish to compare the technique of Section 8.5 with Tukey's "one degree of freedom for non-additivity" (Tukey, 1949) for selecting a re-expression to improve the additivity of a table. The method given here is the natural exploratory analogue of that commonly used technique.

* 8.8   Data Structure

We pause to note the advantages of three-array form as a data structure for median polish. Empty cells, cells with several data values, and unbalanced tables with different numbers of data values in each cell need no special programming. One restriction is that the programs assume that row numbers and column numbers are consecutive and start from 1. If a row or a column is completely missing, the BASIC programs give an error message, and the FORTRAN programs return a zero effect.

In addition, it is possible, through suitable bookkeeping in a driver program, to make some analyses of three-way designs—that is, tables involving a response and three factors. The data structure permits a driver program to maintain three arrays of subscripts—say, row, column, and layer—and pass any *pair* of these arrays to the median-polish program along with the data. This will produce an analysis of the subtable formed by collapsing the table along the un-passed dimension. In this way the "main effects" can be computed easily. (A more sophisticated driver program could use the median-polish routine to fit more complicated models to three- and more-than-three-way tables.)

† 8.9   Algorithms

The programs work by stepping through rows or columns and copying them to a scratch array so that the median can by found. The subscripts of cells from which data values have been taken are preserved so that the newly found row or column median can be subtracted from these cells efficiently. On exit, the residual vector is in exactly the same order as the original data vector and uses the same row and column subscripts. (In the BASIC programs the residuals replace the data vector.)

Comparison values are placed in a vector exactly parallel to, and using the same row and column subscripts as, the data and residuals. This arrangement allows the vector of comparison values and the vector of residuals to be passed as a set of (*x*, *y*) pairs to the *x-y* plot program or to the resistant-line program without having to tell those programs about the subscript arrays.

FORTRAN

The FORTRAN programs for median polish are invoked with the FORTRAN statement

```
CALL MEDPOL(Y, RSUB, CSUB, N, NR, NC, G, RE, CE, RESID, HSTEPS, START,
      SORTY, SUBSAV, NS, ERR)
```

where

Y( )	the data array containing N items;
RSUB( ), CSUB( )	integer arrays containing the N row and column subscripts, respectively, of each element of Y( );

N	is the number of data values;
NR, NC	are the number of rows and columns in the table, respectively;
G	is a REAL variable to return the grand or common level;
RE( ), CE ( )	are REAL arrays dimensioned NR and NC, respectively, to return row and column effects;
RESID( )	is a REAL array to return the N residuals;
HSTEPS	is the number of half-steps to be performed;
START	is a flag (START = 1 tells MEDPOL to start with rows; START = 2, columns);
SORTY( )	is a scratch array for sorting data values;
SUBSAV( )	is an INTEGER scratch array that holds subscripts;
NS	is the dimension of SUBSAV( ) (must be no less than the larger of NR and NC);
ERR	is the error flag, whose values are

	0	normal
	81	the table has a zero dimension (NR = 0 or NC = 0)
	82	no half-steps requested
	83	START not equal to 1 or 2
	85	the table is empty.

Two-way comparison values can be found with a subsequent call to the subroutine TWCVS via the statement

```
CALL TWCVS(RESID, RSUB, CSUB, N, RE, NR, CE, NC, G, CVALS, ERR)
```

where all arguments have the same meanings as described for the subroutine MEDPOL and where

CVALS	is an N-long array in which the comparison values are returned;
ERR	is the error flag, whose values are

	0	normal
	88	common term = 0 (comparison values cannot be computed).

BASIC

The BASIC program for median polish is entered with the data in Y( ) and row and column subscripts in R( ) and C( ), respectively. On entry, N is the length of Y( ), R9 is the number of rows, C9 is the number of columns, and J9 is the number of half-steps to be computed. The version number, V1, has the following effects: V1 = 1 means skip initialization and continue polishing an already polished table, V1 = 2 means initialize and do 4 half-steps starting with rows, V1 ≥ 3 means initialize and do J9 half-steps starting according to the order switch. The order switch, O$, must be set to "ROW" to start the iteration with rows and to "COL" to start the iteration with columns.

On return, Y(1) through Y(N) hold residuals, Y(R8 + 1) through Y(R8 + R9) hold row effects, Y(C8 + 1) through Y(C8 + C9) hold column effects and Y(G8) holds the common or grand effect. The program sets C8 = N, R8 = N + C9, and G8 = N + R9 + C9 + 1. In addition, the subscripts in R( ) and C( ) are extended to indicate that the column effects are in the R9 + 1 row, and the row effects are in the C9 + 1 column. A program (not provided here) to print a table from Y( ), R( ), and C( ) would then place the effects correctly. Placing the effects in a new row and a new column of the data vector is also appropriate for generalizing the program to handle three- or four-way tables.

Reference

Tukey, J.W. 1949. "One Degree of Freedom for Non-additivity." *Biometrics* 5:232–242.

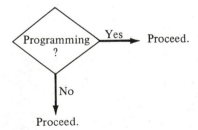

```
5000 REM   MEDIAN POLISH
5010 REM   N=#NUMBERS, R9=#ROWS, C9=#COLS, J9=#ITERATIONS
5020 REM   V1=1: SKIP INITIALIZATION TO DO ADDITIONAL POLISH
5030 REM   V1=2 DEFAULT: 4 HALF-STEPS, STARTS WITH ROWS, FROM SCRATCH.
5040 REM   V1>=3 FROM SCRATCH (INITIALIZES ALL EFFECTS TO ZERO).
5050 REM   O$ = ORDER SWITCH; "ROW" TO START WITH ROWS, "COL" FOR
                                                      COLUMNS.
5060 REM   >>>>DESTROYS ORIGINAL DATA <<<<<<<
5070 REM   RETURNS:   RESIDUALS IN Y(1) THRU Y(N)
5080 REM              ROW EFFECTS IN Y(R8+1) THRU Y(R8+R9)
5090 REM              COL EFFECTS IN Y(C8+1) THRU Y(C8+C9)
5100 REM              GRAND EFFECT IN Y(G8) AND G
5110 REM   WHERE C8=N, R8=N+C9, AND G8=N+R9+C9+1.
5120 REM   THIS PROGRAM USES SPARSE-MATRIX FORM WITH DATA IN Y(), ROW
5130 REM   SUBSCRIPTS IN R(), AND COLUMN SUBSCRIPTS IN C().  IT REQUIRES
5140 REM   N+R9+C9+1 CELLS IN EACH OF X(), Y(), R(), AND C().
5150 REM   THIS PROGRAM CAN HANDLE MISSING CELLS AND UNEQUAL CELL COUNTS.
5160 REM   IF AN ENTIRE ROW OR COLUMN IS MISSING, ITS EFFCT WILL BE ZERO.
5170 REM

5180 LET C8 = N
5190 LET R8 = N + C9
5200 LET G8 = N + R9 + C9 + 1
5210 IF ABS(V1) = 1 THEN 5390

5220 REM   INITIALIZE COLUMN OF ROW EFFECTS

5230 FOR I = 1 TO R9
5240    LET K = R8 + I
5250    LET R(K) = I
5260    LET C(K) = C9 + 1
5270    LET Y(K) = 0
5280 NEXT I

5290 REM    INITIALIZE ROW OF COL EFFECTS

5300 FOR J = 1 TO C9
5310    LET K = C8 + J
5320    LET R(K) = R9 + 1
5330    LET C(K) = J
5340    LET Y(K) = 0
5350 NEXT J
5360 LET R(G8) = R9 + 1
5370 LET C(G8) = C9 + 1
5380 LET Y(G8) = 0

5390 REM   SETUP AND CHECK

5400 IF V1 <> 2 THEN 5430
5410 LET J9 = 4
5420 LET O$ = "ROW"
```

245

```
5430 IF V1 > 0 THEN 5460
5440 PRINT TAB(M0);"HALFSTEPS, 'ROW' OR 'COL'";
5450 INPUT J9,O$
5460 IF O$ = "ROW" THEN 5510
5470 IF O$ = "COL" THEN 5510
5480 PRINT TAB(M0);"SPECIFY 'ROW' OR 'COL'";
5490 INPUT O$
5500 GO TO 5460
5510 IF J9 > 0 THEN 5560
5520 PRINT TAB(M0);J9;" HALF-STEPS IS ILLEGAL."
5530 PRINT TAB(M0);"ENTER #HALF-STEPS BETWEEN 1 AND 12";
5540 INPUT J9
5550 GO TO 5510
5560 IF J9 > 12 THEN 5520
5570 LET J8 = 0
5580 LET N7 = N
5590 IF O$ = "COL" THEN 5930

5600 REM   MEDIAN POLISH FOR ROWS

5610 FOR I = 1 TO R9 + 1
5620    LET L = 0
5630    FOR K = 1 TO N7 + R9 + C9
5640       IF R(K) <> I THEN 5690
5650       IF C(K) > C9 THEN 5690
5660       LET L = L + 1
5670       LET W(L) = Y(K)
5680       LET X(L) = K
5690    NEXT K
5700    IF L > 0 THEN 5770
5710    IF I <= R9 THEN 5740
5720    PRINT TAB(M0);"ALL ROWS EMPTY"
5730    STOP

5740    REM   FLAG EMPTY ROW

5750    LET R(R8 + I) = R9 + 2
5760    GO TO 5900

5770    REM   GET ROW MEDIAN AND ADJUST

5780    LET N = L
5790    GOSUB 1000
5800    LET M5 = FNM((L + 1) / 2)
5810    FOR J = 1 TO L
5820       LET Y(X(J)) = Y(X(J)) - M5
5830    NEXT J
5840    IF I = R9 + 1 THEN 5890

5850    REM   ADD MEDIAN TO ROW EFF

5860    LET Y(R8 + I) = Y(R8 + I) + M5

5870    GO TO 5900
```

```
5880    REM   IF ROW OF COL EFFS, ADD TO GRAND EFF INSTEAD

5890    LET Y(G8) = Y(G8) + M5
5900 NEXT I
5910 LET J8 = J8 + 1
5920 IF J8 >= J9 THEN 6250

5930 REM   MEDIAN POLISH FOR COLUMNS

5940 FOR J = 1 TO C9 + 1
5950    LET L = 0
5960    FOR K = 1 TO N7 + R9 + C9
5970      IF C(K) <> J THEN 6020
5980      IF R(K) > R9 THEN 6020
5990      LET L = L + 1
6000      LET W(L) = Y(K)
6010      LET X(L) = K
6020    NEXT K
6030    IF L > 0 THEN 6100
6040    IF J <= C9 THEN 6070
6050    PRINT TAB(M0);"ALL COLS EMPTY"
6060    STOP

6070    REM   MARK MISSING COLUMN

6080    LET C(C8 + J) = C9 + 2
6090    GO TO 6220
6100    LET N = L
6110    GOSUB 1000
6120    LET M5 = FNM((L + 1) / 2)
6130    FOR I = 1 TO L
6140      LET Y(X(I)) = Y(X(I)) - M5
6150    NEXT I
6160    IF J = C9 + 1 THEN 6200

6170    REM   ADD MEDIAN TO COL EFF

6180    LET Y(C8 + J) = Y(C8 + J) + M5
6190    GO TO 6220

6200    REM   IF COL OF ROW EFFS, ADD TO GRAND EFF

6210    LET Y(G8) = Y(G8) + M5
6220 NEXT J
6230 LET J8 = J8 + 1
6240 IF J8 < J9 THEN 5600

6250 REM   DONE

6260 LET N = N7
```

```
6270 REM   MAKE SUBSCRIPTS OF MISSING EFFECTS LEGAL AGAIN

6280 FOR I = 1 TO R9
6290    IF R(R8 + I) <= R9 + 1 THEN 6320
6300    LET R(R8 + I) = I
6310    LET Y(R8 + I) = 0
6320 NEXT I
6330 FOR J = 1 TO C9
6340    IF C(C8 + J) <= C9 + 1 THEN 6370
6350    LET C(C8 + J) = J
6360    LET Y(C8 + J) = 0
6370 NEXT J
6380 LET N = N7
6390 LET G = Y(G8)
6400 IF G <> 0 THEN 6430
6410 PRINT TAB(M0);"GRAND EFFECT=0, CANNOT COMPUTE COMPARISON VALUES"
6420 GO TO 6460
6430 FOR K = 1 TO N
6440    LET X(K) = (Y(R8 + R(K)) * Y(C8 + C(K))) / G
6450 NEXT K
6460 RETURN
6470 END
```

```
      SUBROUTINE MEDPOL(Y, RSUB, CSUB, N, NR, NC, G, RE, CE, RESID,
     1 HSTEPS, START, SORTY, SUBSAV, NS, ERR)
C
      INTEGER N, NR, NC, HSTEPS, START, NS, ERR
      INTEGER RSUB(N), CSUB(N), SUBSAV(NS)
      REAL Y(N), G, RE(NR), CE(NC), RESID(N), SORTY(N)
C
C  ANALYZE THE TWO-WAY TABLE IN Y() BY MEDIAN POLISH.
C  THE TABLE HAS  NR  ROWS AND  NC  COLUMNS, BUT IS REPRESENTED IN
C  THREE ARRAYS:  RSUB(I)  AND  CSUB(I)  CONTAIN THE (ROW, COL)
C  SUBSCRIPTS OF THE DATA VALUE IN  Y(I).   THIS PERMITS MULTIPLE
C  OBSERVATIONS IN A CELL OF THE TABLE OR A COMPLETELY MISSING CELL
C  AND MAKES MANY MANIPULATIONS EASIER.
C      ON EXIT, Y()  IS UNCHANGED, G  IS THE GENERAL TYPICAL (OR
C  COMMON) VALUE,  RE()  AND  CE()  ARE THE ROW EFFECTS AND COLUMN
C  EFFECTS, RESPECTIVELY, AND  RESID()  IS THE TWO-WAY TABLE OF
C  RESIDUALS IN THE SAME FORMAT AS  Y() (USING RSUB() AND  CSUB()).
C  THE RESIDUALS ARE DEFINED BY
C
C          RESID(I, J) = Y(I, J) - G - RE(I) - CE(J)
C
C  AND ACTUALLY STRUCTURED AS
C
C          RESID(K) = Y(K) - G - RE(RSUB(K)) - CE(CSUB(K))
C
C  ANY ROW OR COLUMN FOUND TO BE ENTIRELY MISSING IN THE ORIGINAL
C  DATA WILL HAVE ITS EFFECT SET TO ZERO ON EXIT.
C
C  THE INPUT PARAMETERS  HSTEPS  AND  START  CONTROL THE
C  ITERATION PROCESS.  HSTEPS  IS THE NUMBER OF HALF-STEPS TO BE
C  PERFORMED, AND  START  DETERMINES WHETHER THE FIRST STEP
C  OPERATES ON ROWS (START = 1) OR ON COLUMNS (START = 2).
C  THE INTEGER VECTOR  SUBSAV()  IS USED TO STORE SUBSCRIPTS
C  TEMPORARILY.  ITS DIMENSION,  NS, MUST BE AT LEAST AS LARGE AS
C  THE LARGER OF  NR  AND  NC.
C
C
C  FUNCTION
C
      REAL MEDIAN
C
C  LOCAL VARIABLES
C
      INTEGER I, J, K, L, IROW, ICOL, ISTEP
      REAL REFF, CEFF, EMPTY
      DATA EMPTY/987.654/
C
C  EMPTY  IS AN INTERNAL FLAG USED TO MARK EMPTY ROWS OR COLUMNS.
C  THE VALUE USED HERE IS ARBITRARY.
C
```

249

```
C
C   CHECK VALIDITY OF INPUT
C
      IF(NR .GT. 0 .AND. NC .GT. 0) GO TO 4
      ERR = 81
      GO TO 999
    4 IF(HSTEPS .GT. 0) GO TO 8
      ERR = 82
      GO TO 999
    8 IF(START .EQ. 1 .OR. START .EQ. 2) GO TO 10
      ERR = 83
      GO TO 999
C
C   INITIALIZE RE AND CE TO ZERO, RESID TO Y, AND ISTEP TO 0.
C
   10 DO 20 I = 1, NR
         RE(I) = 0.0
   20 CONTINUE
      DO 30 J = 1, NC
        CE(J) = 0.0
   30 CONTINUE
      DO 40 K = 1, N
         RESID(K) = Y(K)
   40 CONTINUE
      ISTEP = 0
C
C   BEGIN ON ROWS IF START=1, ELSE BEGIN ON COLUMNS.
C
      IF(START .EQ. 2) GO TO 130
C
C   FIND ELEMENTS OF EACH ROW,  FIND ROW MEDIANS, ADD THEM TO ROW
C   EFFECTS, AND SUBTRACT THEM FROM PREVIOUS RESIDUALS.
C
   50 IF(ISTEP .GE. HSTEPS) GO TO 210
      DO 120 IROW = 1, NR
        IF(RE(IROW) .EQ. EMPTY) GO TO 120
        L = 0
C
C   SEARCH FOR ANY MATCHING ROW SUBSCRIPT
C
        DO 60 K = 1, N
          IF(RSUB(K) .NE. IROW) GO TO 60
          L = L+1
          SORTY(L) =  RESID(K)
          SUBSAV(L) = K
   60     CONTINUE
        IF(L .GT. 0) GO TO 70
```

```
C
C   NO DATA IN THIS ROW, MARK THE ROW EMPTY TO AVOID FUTURE SEARCHES
C
         RE(IROW) = EMPTY
         GO TO 120
   70    IF(L .GT. 1) GO TO 80
         REFF = SORTY(1)
         GO TO 100
   80    IF(L .EQ. 2) GO TO 90
         CALL SORT(SORTY, L, ERR)
         IF(ERR .NE. 0) GO TO 999
   90    REFF = MEDIAN(SORTY, L)
C
C   ADJUST FOR ROW EFFECT NOW IN REFF
C
  100    RE(IROW) = RE(IROW) + REFF
         DO 110 I = 1, L
            J = SUBSAV(I)
            RESID(J) = RESID(J) - REFF
  110    CONTINUE
  120 CONTINUE
      ISTEP = ISTEP + 1
C
C   FIND ELEMENTS OF EACH COLUMN, FIND COLUMN MEDIANS, ADD THEM TO
C   COLUMN EFFECTS, AND SUBTRACT THEM FROM PREVIOUS RESIDUALS.
C
  130 IF(ISTEP .GE. HSTEPS) GO TO 210
      DO 200 ICOL = 1, NC
         IF(CE(ICOL) .EQ. EMPTY) GO TO 200
         L = 0
C
C   SEARCH FOR ANY MATCHING COLUMN SUBSCRIPT
C
         DO 140 K = 1, N
            IF(CSUB(K) .NE. ICOL) GO TO 140
            L = L+1
            SORTY(L) = RESID(K)
            SUBSAV(L) = K
  140    CONTINUE
         IF(L .GT. 0) GO TO 150
C
C NO DATA IN THIS COLUMN, MARK IT EMPTY TO AVOID FUTURE SEARCHES
C
         CE(ICOL) = EMPTY
         GO TO 200
  150    IF(L .GT. 1) GO TO 160
         CEFF = SORTY(1)
         GO TO 180
  160    IF(L .EQ. 2) GO TO 170
         CALL SORT(SORTY, L, ERR)
         IF(ERR .NE. 0) GO TO 999
  170    CEFF = MEDIAN(SORTY, L)
```

```
C
C   ADJUST FOR COLUMN EFFECT NOW IN CEFF.
C
   180    CE(ICOL) = CE(ICOL) + CEFF
          DO 190 I = 1, L
             J = SUBSAV(I)
             RESID(J) = RESID(J) - CEFF
   190    CONTINUE
   200 CONTINUE
          ISTEP = ISTEP+1
          GO TO 50
C
C   NOW CENTER ROW EFFECTS AND COLUMN EFFECTS TO HAVE MEDIAN ZERO,
C   AND COMBINE THE CONTRIBUTIONS TO THE COMMON VALUE.
C
   210 L = 0
       DO 220 I = 1, NR
          IF(RE(I) .EQ. EMPTY) GO TO 220
          L = L+1
          SORTY(L) = RE(I)
   220 CONTINUE
       IF(L .NE. 0) GO TO 230
       ERR = 85
       GO TO 999
   230 CALL SORT(SORTY, L, ERR)
       IF(ERR .NE. 0) GO TO 999
       G = MEDIAN(SORTY, L)
       DO 240 I = 1, NR
          IF(RE(I) .NE. EMPTY) RE(I) = RE(I) - G
C
C   RETURN ZERO FOR EFFECT OF EMPTY ROW
C
          IF(RE(I) .EQ. EMPTY) RE(I) = 0.0
   240 CONTINUE
       L = 0
       DO 250 J = 1, NC
          IF(CE(J) .EQ. EMPTY) GO TO 250
          L = L+1
          SORTY(L) = CE(J)
   250 CONTINUE
       IF(L .NE. 0) GO TO 260
       ERR = 85
       GO TO 999
   260 CALL SORT(SORTY, L, ERR)
       IF(ERR .NE. 0) GO TO 999
       CEFF = MEDIAN(SORTY, L)
       G = G+CEFF
       DO 270 J = 1, NC
          IF(CE(J) .NE. EMPTY) CE(J) = CE(J) - CEFF
```

```
C
C   RETURN ZERO FOR EFFECT OF EMPTY COLS
C
          IF(CE(J) .EQ. EMPTY) CE(J) = 0.0
  270 CONTINUE
C
  999 RETURN
      END


      SUBROUTINE TWCVS(RSUB, CSUB, N, RE, NR, CE, NC, G, CVALS,
     1 ERR)
C
      INTEGER NR, NC, N, ERR
      INTEGER RSUB(N), CSUB(N)
      REAL RE(NR), CE(NC), G, CVALS(N)
C
C   CALCULATES THE  COMPARISON VALUES  FOR A TWO-WAY
C   TABLE.   THE FIT ON WHICH THESE ARE BASED CONSISTS OF THE
C   ROW EFFECTS,  RE(1),...,RE(NR) , THE COLUMN EFFECTS,
C   CE(1),...,CE(NC) , AND THE COMMON VALUE,  G .  BY
C   DEFINITION, THE COMPARISON VALUE FOR CELL (I,J) IS
C
C                 RE(I) * CE(J) / G        .
C
C   CVALS() IS INDEXED BY THE ROW AND COLUMN SUBSCRIPTS
C   FOUND IN THE CORRESPONDING LOCATIONS IN RSUB() AND CSUB().
C   THIS SUBROUTINE IDENTIFIES THE ROW AND COLUMN EFFECTS ASSOCIATED
C   WITH EACH RESIDUAL AND PUTS
C   THE CORRESPONDING COMPARISON VALUES IN CVALS().
C
C
C   LOCAL VARIABLES
C
      INTEGER I, J, K
C
      IF(NR .GT. 0 .AND. NC .GT. 0) GO TO 10
        ERR = 81
        GO TO 999
   10 IF(G .NE. 0.0) GO TO 30
        ERR = 88
        GO TO 999
C
   30 DO 50 K = 1,N
        I = RSUB(K)
        J = CSUB(K)
        CVALS(K) = RE(I) * CE(J) / G
   50 CONTINUE
  999 RETURN
      END
```

Chapter 9

Rootograms

bins

Batches of data are sometimes recorded by splitting the range of possible values into intervals, or *bins,* and simply counting the data values that fall into each bin. In a large batch, lack of room to construct a stem-and-leaf display would lead us to use bins. If we had 500 data values, we would usually record how many values fall on each line of the display instead of showing a leaf for each data value.

Some variables almost always take this form. For example, ages of adults seldom appear in more detail than the year (for most purposes five-year or ten-year intervals are standard), so it is common to report age data as counts of people at each age or in each age category. When the individual data value is a count—especially a small count—there are often many repeated values, and it is easiest to record the number of times each possible value occurs. For example, from data on the number of traffic tickets that individual drivers received in one year, we would record how many drivers received zero tickets, how many received one ticket, and so on.

This chapter shows how to display such batches effectively, how to compare them to standard shapes, and what residuals to calculate in these comparisons. The exploratory techniques are known as the *rootogram*—for basic display—and the *suspended rootogram*—for comparisons and residuals.

255

Almost all introductory statistics texts discuss the "normal" distribution, and most imply that it is common for data in general—and especially for data reported by bins—to be well-described by the "normal" shape. A little experience exploring data shows that the "normal" distribution is, in fact, rather rare. (This is one reason the distribution has been called "Gaussian" in this book.) Nevertheless, the Gaussian shape—a symmetric bell shape (Exhibit 9–1)—is a useful standard against which to compare the distribution of data values in a batch. We do often observe many data values piling up in the middle bins and fewer values in bins further from the middle. However, we also often see skewed shapes or unusually full or empty bins. The methods discussed in this chapter make it easy to find these and other deviations from the Gaussian standard.

The exploratory methods in earlier chapters required no background in mathematics or statistics. While the principle of the suspended rootogram is easy to understand (compare Exhibits 9–12, 9–13, and 9–14), you will need to know a little basic statistics to understand how to make one. Primarily, you should be acquainted with the Gaussian (or normal) distribution and with the idea that area under a density curve (the "bell-shaped" curve, for the Gaussian distribution) can be interpreted as a probability. Most statistics texts provide a table of these probabilities. We do not need such a table in this chapter, but you may be able to use one in approximately checking some of the calculations. If you lack this background in statistics, you can still read this chapter, and you will certainly be able to use suspended rootograms, but you may want to read lightly over the sections that discuss the method in detail. Even readers who have the necessary background will still have to accept a few

Exhibit 9–1   The Frequency Curve of the Standard Gaussian Distribution

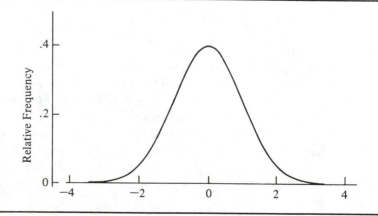

statements without rigorous justification. Readers with more extensive statistical background will find greater detail and relevant references in Section 9.7.

9.1 Histograms and the Area Principle

Histograms

histogram

If we want to see only skeletal detail in a stem-and-leaf display, we can trace the outline of the lines of leaves. The result is a *histogram,* and it is customarily presented with the data axis horizontal and the bars vertical. Exhibit 9–2 shows the histogram obtained by tracing a stem-and-leaf display for the precipitation pH data in Exhibit 1–1. Here each line of the stem-and-leaf display defines a bin.

Instead of a stem-and-leaf display, the data might take the form of a set of counts as in Exhibit 9–3, which lists the intervals of pH value and the number of data values that belong to each of them. (Other sets of intervals are possible; Exhibit 9–3 simply uses the ones established in the stem-and-leaf display in Exhibit 1–2.) Another name for data in the form of Exhibit 9–3 is *frequency distribution;* the tabulation shows how often the data values fall in each interval.

frequency distribution

Exhibit 9–2   A Histogram for the Precipitation pH Data

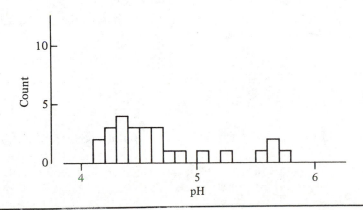

Exhibit 9–3   A Frequency Distribution for the Precipitation pH Data of Exhibit 1–1

pH	Number of Precipitation Events
4.10 – 4.19	2
4.20 – 4.29	3
4.30 – 4.39	4
4.40 – 4.49	3
4.50 – 4.59	3
4.60 – 4.69	3
4.70 – 4.79	1
4.80 – 4.89	1
4.90 – 4.99	0
5.00 – 5.09	1
5.10 – 5.19	0
5.20 – 5.29	1
5.30 – 5.39	0
5.40 – 5.49	0
5.50 – 5.59	1
5.60 – 5.69	2
5.70 – 5.79	1
	26

The Area Principle

area principle

To make a histogram for a large batch, where using digits for leaves in a stem-and-leaf display would require too much space, we need only represent each data value by the same amount of area. This is the *area principle.* This principle is important in many displays because visual impact is generally proportional to area.

Equal-Width Bins

In the simplest situation, all the bins span equal ranges of data values. Exhibit 9–3, for example, uses bins 0.10 pH-units wide for the precipitation pH data. When all the bins have the same width, a histogram of the data will have bars of equal physical width. Then, to make impact proportional to count, we simply give each bar of the histogram a height that is a constant multiple of the count—that is, the number of data values—in its bin.

Exhibit 9–4 shows a larger example, the chest measurements of 5738 Scottish militiamen. The data have some historical significance because they figured in a 19th-century discussion of the distribution of various human characteristics. The source for these data is an 1846 book by the Belgian statistician Adolphe Quetelet, but the data were first published about thirty years earlier. These measurements were recorded in one-inch intervals; so all the bins have the same width—one inch of chest measurement, centered at a whole number of inches. Exhibit 9–5 shows a histogram based on Exhibit 9–4. The constant of proportionality relating the height of each bar to the count in the corresponding bin affects only the scale of the vertical axis; so we do not have to calculate this constant explicitly. In Exhibit 9–5 we see a fairly well-behaved shape: The middle bars are longest, and the bars regularly become shorter as we move toward either end of the batch. (In Section 9.4 we

---

Exhibit 9–4   Chest Measurements of 5738 Scottish Militiamen

Chest (in.)	Count
33	3
34	18
35	81
36	185
37	420
38	749
39	1073
40	1079
41	934
42	658
43	370
44	92
45	50
46	21
47	4
48	1
	5738

Source: Data from A. Quetelet, *Lettres à S.A.R. le Duc Régnant de Saxe-Cobourg et Gotha, sur la Théorie des Probabilités, Appliquée aux Sciences Morales et Politiques.* (Brussels: M. Hayez, 1846) p. 400.

Exhibit 9–5   Histogram for the Chest Measurement Data in Exhibit 9–4

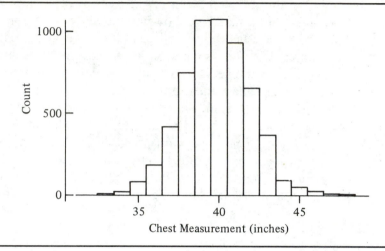

will summarize this shape and examine how the data depart from the summary.)

Unequal-Width Bins

When the bins do not all have the same width, we must make the physical width of their histogram bars reflect the bin widths and take these different widths into account in order to preserve the area principle. Fortunately, we need only make the height of each bar proportional to the count in its bin divided by the width of that bin. A little more detailed discussion shows how this process works.

bin boundaries   We assume that the data set consists of a set of **bin boundaries,**

$$x_0, x_1, \ldots, x_k,$$

bin counts   and a set of **bin counts,**

$$n_0, n_1, \ldots, n_k, n_{k+1},$$

where n_i is the count in the bin whose right-hand boundary is x_i. Thus, the first and last bins are unbounded on one side: n_0 data values are below x_0 and n_{k+1} data values are above x_k. If unbounded bins do not arise, then $n_0 = n_{k+1} = 0$, and we do not have to worry about the problem of what bin width to use for the unbounded bins. When unbounded bins do arise, we must take care to depict them fairly in any display of the data. The total count in the batch is $N = n_0 + n_1 + \ldots + n_{k+1}$. The bin widths are the differences between successive x_i; that is, $w_1 = x_1 - x_0, \ldots, w_k = x_k - x_{k-1}$.

When the bin widths vary, the widths of the histogram bars will also vary. We construct a histogram by choosing the width of each bar proportional to the bin width and then choosing the height of each bar so that the area of the bar is proportional to the bin count. These proportionality constants affect only the scaling of the axes, and we omit them from the derivations. Thus, if *bin width* the **bin width** is $w_i = x_i - x_{i-1}$ and the bar height is to be d_i, we take

$$d_i = n_i/w_i.$$

As defined in this equation, d_i gives the density of data values in the interval—that is, the number of data values per unit of bin width.

In a discussion involving nutrition, Huffman, Chowdhury, and Mosley (1979) present data on two samples of women in Bangladesh. The height data for one of their samples are shown in Exhibit 9–6, along with the width of each bin and the calculated bar height, d_i. This set of data has an unbounded bin at each end. Because we cannot be sure whether these end bins represent

Exhibit 9–6 A Frequency Distribution and the Histogram Calculations for the Heights of 1243 Women in Bangladesh

Height (cm)	Number of Women (n_i)	Bin Width (w_i)	Count per Width (d_i)
< 140.0	71	?	?
140.0 – 142.9	137	3	45.67
143.0 – 144.9	154	2	77.00
145.0 – 146.9	199	2	99.50
147.0 – 149.9	279	3	93.00
150.0 – 152.9	221	3	73.67
153.0 – 154.9	94	2	47.00
155.0 – 156.9	51	2	25.50
> 156.9	37	?	?
	1243		

Source: S.L. Huffman, A.K.M. Alauddin Chowdhury, and W.H. Mosley, "Difference between Postpartum and Nutritional Amenorrhea (reply to Frisch and McArthur)," *Science* 203 (1979):922–923. Copyright 1979 by the American Association for the Advancement of Science. Reprinted by permission.

intervals of width 2 or 3 or some other value, we do not attempt to find the height of a histogram bar for them. (This will not, however, prevent us from comparing this frequency distribution to a Gaussian distribution and calculating a residual in each bin, as we will see in Section 9.4.) Exhibit 9–7 shows the histogram. Again, as in Exhibit 9–5, the pattern of bars looks quite regular.

The process of constructing a histogram involves nothing more than the simple calculations that we have made so far. When we examine a set of data closely, however, we often want to go beyond the histogram. After we pick out the major features in the histogram, as we would do for a stem-and-leaf display, we are then ready to compare the data to some standard of behavior and look further for patterns in the residuals.

9.2 Comparisons and Residuals

When we compare a histogram to some expected pattern of behavior, we must accept variability among data sets and among their histograms. If we studied a

Exhibit 9–7   Histogram for the Heights of Bangladesh Women (Data from Exhibit 9–6)

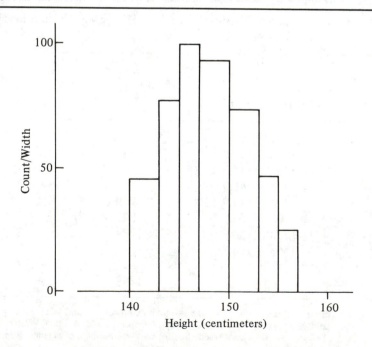

large number of histograms from closely related sets of data—for example, many samples of women's heights in Bangladesh—we would generally find that bar height varies more in bins with long bars than in bins with short bars. Put in terms of the counts in the frequency distribution, the variability of the counts increases as their typical size increases. This is hardly surprising. A count that is typically 2 might often come out 1 or 0 or 3 or 4 in an observed frequency distribution, but it would rarely come out 10. However, if the count is typically 100, observed values of 90 or 110 would be quite common. Thus, when we make direct comparisons and use residuals to look closer at patterns of deviation, we must take into account the fact that variability is not constant from one bin to another.

A re-expression can approximately remove the tendency for the variability of a count to increase with its typical size. The most helpful re-expression is a familiar one: the square root. In addition to its helpful quality of stabilizing variability, the square-root re-expression for counts has some theoretical justifications. We consider some of these justifications in the next section (and in Section 9.7).

9.3 Rootograms

rootogram

When we apply the square-root re-expression to a histogram, we obtain a *rootogram*. The bin widths (w_i) have not changed; so we keep the same bar widths as in the histogram, but we now use $\sqrt{d_i}$ as the height of the bar for bin *i*. The chest measurement data of Exhibit 9–4 provide a straightforward example. Exhibit 9–8 gives the square-root calculations, and Exhibit 9–9 shows the rootogram. In the rootogram we see a regular pattern, just as we found in the histogram (Exhibit 9–5). When we compare Exhibits 9–5 and 9–9 more closely, we find that the rootogram looks much more regular—almost inviting us to drape a curve over it—primarily because the square-root re-expression has more impact on the longer bars in the middle than on the shorter bars toward the ends. Just as we saw in earlier chapters, a suitable re-expression can make data more regular and easier to look at.

Note that in using a rootogram we have abandoned the area principle—area is no longer proportional to count. As we move from display to analysis, and so from examining the raw data to fitting a shape and examining the residuals, it will be more important to stabilize the variability of fluctuations than to picture the raw counts directly in terms of area.

Exhibit 9–8   Rootogram Calculations for the Chest Measurement Data of Exhibit 9–4 ($w_i = 1$ for all bins)

Chest (in.)	Count ($=d_i$)	$\sqrt{d_i}$
33	3	1.73
34	18	4.24
35	81	9.00
36	185	13.60
37	420	20.49
38	749	27.37
39	1073	32.76
40	1079	32.85
41	934	30.56
42	658	25.65
43	370	19.24
44	92	9.59
45	50	7.07
46	21	4.58
47	4	2.00
48	1	1.00

Exhibit 9–9   Rootogram for the Chest Measurement Data (All bins have width = 1.)

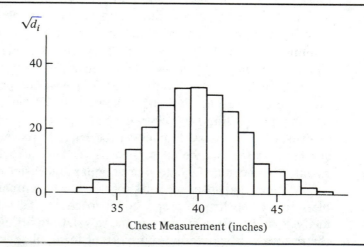

Double-Root Residuals

When we compare a set of observed counts to the corresponding fitted counts, we want to calculate and examine residuals. We could simply subtract fitted from observed, but this would do nothing to make fluctuations roughly the same size across all bins. Therefore, we will work with both observed counts and fitted counts in a square-root scale. We can form residuals in this scale in such a way that they behave approximately like observations from a standard Gaussian distribution and hence are easy to interpret.

We could take

$$\sqrt{\text{observed}} - \sqrt{\text{fitted}}$$

as the residual, but a slightly different re-expression avoids some difficulties with small counts. First we replace the observed count by

$$\sqrt{2 + 4\,(\text{observed})} \qquad \text{if observed} \neq 0$$

$$1 \qquad \text{if observed} = 0$$

and we replace the fitted count by

$$\sqrt{1 + 4\,(\text{fitted})}.$$

double-root residual   Then we define the *double-root residual* (*DRR*) as the difference between these two:

$$DRR = \sqrt{2 + 4\,(\text{observed})} - \sqrt{1 + 4\,(\text{fitted})} \qquad \text{if observed} \neq 0$$

$$= \quad 1 - \sqrt{1 + 4\,(\text{fitted})} \qquad \text{if observed} = 0.$$

These square-root re-expressions have the name "double root" because they are close to two times the usual square root. We will soon see that, as a result, the double-root residuals have an especially convenient scale.

The constants that have been added, 2 for observed and 1 for fitted, help to relieve the compression imposed on small counts by the restriction that counts always be greater than or equal to zero. Because fitted counts are almost always greater than zero—although they can sometimes be small fractions—we need not add as large a constant to fitted counts: 1 will do instead of 2. (Section 9.7 provides some further background on double roots.)

Throughout this section we treat the fitted values as given; nothing has been said about how to calculate them because we would first need to choose a specific model for a frequency distribution. (Section 9.4 describes one technique for fitting a comparison curve to a histogram.)

Pollard (1973) examined the number of points scored per game by individual teams in the 1967 U.S. collegiate football season and then grouped the scores so that each bin corresponds, as nearly as possible, to an exact number of touchdowns (one touchdown = 6 points). The grouped data are in Exhibit 9–10. Pollard devised a model for these data that gives the fitted counts shown in Exhibit 9–10. The corresponding double-root residuals are computed in the last three columns of Exhibit 9–10. The last group, labeled "74 & up," actually contains three scores of 77 and one each of 75, 81, and 90; so it combines what could have been three bins (74–80, 81–87, and 88–94). The practice of combining bins or intervals in order to avoid working with small fitted counts is widespread but is unnecessary when we use double-root residuals.

None of the double-root residuals in the last column of Exhibit 9–10 seem especially large, but we must judge the size of such residuals according to

Exhibit 9–10   U.S. Collegiate Football Scores, with Fitted Counts and Double-Root Residuals

Number of Points per Game	Number of Games		$\sqrt{2 + 4 \text{ (observed)}}$	$\sqrt{1 + 4 \text{ (fitted)}}$	DRR
	Observed	Fitted			
0 – 5	272	278.7	33.02	33.40	−0.39
6 – 11	485	490.2	44.07	44.29	−0.22
12 – 17	537	509.1	46.37	45.14	1.23
18 – 24	407	406.6	40.37	40.34	0.03
25 – 31	258	275.9	32.16	33.24	−1.08
32 – 38	157	167.3	25.10	25.89	−0.79
39 – 45	101	93.5	20.15	19.36	0.78
46 – 52	57	49.0	15.17	14.04	1.13
53 – 59	23	24.4	9.70	9.93	−0.23
60 – 66	8	11.7	5.83	6.91	−1.08
67 – 73	5	5.4	4.69	4.75	−0.06
74 & up	6	4.3	5.10	4.27	0.83
	$N = 2316$				

Source: Data from R. Pollard, "Collegiate Football Scores and the Negative Binomial Distribution," *Journal of the American Statistical Association* 68 (1973):351–352. Reprinted by permission.

some standard. Usually (as in Chapters 5 and 8) we examine residuals as a batch to get an indication of their typical size and to identify any large residuals. Double-root residuals, however, come with a built-in standard of size. When the model fits the data well, an individual double-root residual behaves approximately like an observation from the Gaussian (or normal) distribution with mean 0 and variance 1. Thus, nearly 95 percent of the time a *DRR* should be between -1.96 and $+1.96$. These limits can be found from the table of the "normal" distribution given in most statistics texts. It is convenient to define a large *DRR* as one below -2 or above $+2$. When the fitted count is less than 1.0, the *DRR* may be less like a Gaussian observation; so it may be wise to look more closely at any *DRR* below -1.5 or above $+1.5$.

By these standards, Pollard's model fits quite well (perhaps too well): The largest *DRR* is 1.23. It would be interesting to fit the same model to data from other collegiate football seasons.

In this section we have seen that rootograms stabilize the variability from bar to bar while preserving the form of a histogram and that double-root residuals provide an effective numerical way to compare data with fit. We now turn to one technique for fitting smooth curves to counts in bins.

9.4 Fitting a Gaussian Comparison Curve

When a histogram summarizes a large batch in terms of a set of bins, it is common practice to superimpose a smooth frequency curve on the histogram. The most common curve for this purpose is the one belonging to the Gaussian distribution. Its *standard* form (mean = 0, variance = 1) is given for all values of *z*, positive and negative alike, by the mathematical function

$$f(z) = \frac{1}{\sqrt{2\pi}} e^{-z^2/2},$$

where π and e are common mathematical constants: $\pi \simeq 3.14159$, $e \simeq 2.71828$. A graph of this function against *z* follows the bell shape shown in Exhibit 9–1.

To match this standard curve to a batch of data, we can slide it until its center matches the middle of the batch and stretch it (or compress it) uniformly until its hinges match the hinges of the batch. Because the area beneath $f(z)$ is 1, we must also multiply by N so that the curve represents the same total count as the batch. The result is the curve

$$\frac{N}{s} f\left(\frac{x - m}{s}\right) = \frac{N}{s\sqrt{2\pi}} e^{-(x-m)^2/(2s^2)},$$

whose mean, m, and standard deviation, s, can be calculated from the hinges of the data. Specifically, if H_L and H_U are the lower and upper hinges, respectively, we take

$$m = \tfrac{1}{2}(H_L + H_U)$$

and

$$s = (H_U - H_L)/1.349,$$

because any Gaussian distribution has its hinges at $m - 0.6745s$ and $m + 0.6745s$ and thus has an H-spr of $2 \times 0.6745s = 1.349s$. We could use the data in other ways to calculate m and s. For example, we might (as is often done) use the sample mean for m and the sample standard deviation for s. The hinges, however, are resistant to the ill effects of outliers and are often available in exploratory summaries such as the letter-value display. When we cannot obtain the hinges from the complete data, we may still be able to estimate them by interpolation.

Interpolated Hinges

When we must work from the bin boundaries,

$$x_0, x_1, \ldots, x_k,$$

and the bin counts,

$$n_0, n_1, \ldots, n_k, n_{k+1},$$

as in Section 9.1, we generally do not know the hinges of the data exactly. Nevertheless, we can easily find the bins that contain the two hinges and then estimate a value for each hinge by interpolation. From the total count, N, we know (Section 2.1) that the depth of the hinges is given by $d(H) = [(N + 1)/2 = 1]/2$. The bins at which the sums of the bin counts (the n_i), summing in from each end, first exceed or equal $d(H)$ are the bins that contain the hinges.

Let us suppose that the lower hinge lies in the bin whose boundaries are x_{L-1} and x_L and whose observed count is n_L. Then we interpolate by treating the n_L data values in the bin as if they were spread evenly across the width of the bin. More specifically, we act as if the bin is divided into n_L equal subintervals of width w_L/n_L, each with a data value at its center. (Recall that $w_L = x_L - x_{L-1}$.) Thus, the leftmost spread-out value falls at

$$x_{L-1} + \frac{0.5w_L}{n_L},$$

the next value comes at

$$x_{L-1} + \frac{1.5w_L}{n_L},$$

and so on. Thus if the depth of the hinge is $d(H)$, we place the interpolated lower hinge at

$$x_{L-1} + \frac{d(H) - (n_0 + \ldots + n_{L-1}) - 0.5}{n_L} w_L.$$

For the chest measurement data in Exhibit 9–4, we have $N = 5738$, so

$$d(H) = \frac{[(5739)/2 + 1]}{2} = 1435.$$

Summing the bin counts from the low end of the frequency distribution, we find that

$$n_0 + \ldots + n_5 = 0 + 3 + \ldots + 420 = 707$$

and

$$n_0 + \ldots + n_6 = 0 + 3 + \ldots + 749 - 1456.$$

Thus, because $w_i = 1$ for all the bins, we estimate the lower hinge as

$$37.5 + \frac{1435 - 707 - 0.5}{749} \times 1 = 38.471.$$

Similarly, if the upper hinge lies in the bin whose boundaries are x_{U-1} and x_U—that is, $n_{U+1} + \ldots + n_{k+1} < d(H) \leq n_U + \ldots + n_{k+1}$—we place the interpolated upper hinge at

$$x_U - \frac{d(H) - (n_{U+1} + \ldots + n_{k+1}) - 0.5}{n_U} w_U.$$

Warning: If either hinge lies in a half-open bin—that is, to the left of x_0 or to the right of x_k—we will be unable to interpolate and hence unable to fit the comparison curve from the interpolated hinges. (The computer programs in this chapter check for this unlikely possibility and indicate an error condition if it occurs.) Such a situation may require a re-expression of the data.

Fitted Counts

Finally, from the fitted comparison curve, we must obtain a fitted count for each bin. The fitted count is just the area beneath the fitted curve, $(N/s) \times f((x - m)/s)$, between the bin boundaries. We could approximate this area fairly closely by multiplying the bin width by the height of the curve at the center of the bin, but we would have difficulty with the half-open bins (which can have appreciable fitted counts even when their observed counts are zero). Thus we employ, instead, the cumulative distribution function, F, for the standard Gaussian distribution.

cumulative distribution function   The ***cumulative distribution function*** tells how much probability lies to the left of any given value on the scale of the data. When we fit a Gaussian shape, $F(z)$ is the amount of probability to the left of z in the standard Gaussian distribution. For the fitted Gaussian comparison curve,

$$N \times F\left(\frac{x_i - m}{s}\right)$$

is the total fitted count to the left of x_i. We can thus begin with the left half-open bin and calculate its fitted count, $\hat{n}_0$, from

$$\hat{n}_0 = N \times F\left(\frac{x_0 - m}{s}\right)$$

and continue by calculating

$$\hat{n}_1 = N\left(F\left(\frac{x_1 - m}{s}\right) - F\left(\frac{x_0 - m}{s}\right)\right)$$

and so on. In general,

$$\hat{n}_i = N\left(F\left(\frac{x_i - m}{s}\right) - F\left(\frac{x_{i-1} - m}{s}\right)\right)$$

except for the right half-open bin:

$$\hat{n}_{k+1} = N\left(1 - F\left(\frac{x_k - m}{s}\right)\right).$$

If we wish, we can sketch in the comparison curve as a background for a rootogram, but we calculate double-root residuals from the n_i and the $\hat{n}_i$.

The standard Gaussian cumulative function, F, has no simple formula like that given earlier for the density function, f. Good approximations for F are available, however, for computers or calculators. The programs at the end of this chapter use a reasonably accurate simple approximation developed by Derenzo (1977) for use on hand-held calculators: If $|z| \le 5.5$, $f(z)$ is approximated by setting $v = |z|$, calculating

$$p = \exp\left\{-\frac{((83v + 351)v + 562)v}{703 + 165v}\right\}$$

and returning

$$F(z) = \tfrac{1}{2}p \qquad \text{if } z \le 0$$

$$F(z) = 1 - \tfrac{1}{2}p \qquad \text{if } z > 0.$$

When $|z| > 5.5$, the FORTRAN program uses another approximation from Derenzo, while the BASIC program sets p to zero in the preceding equation for $F(z)$. Because $|z| > 5.5$ corresponds to a probability smaller than 1/10,000,000, this difference between the programs is of no practical consequence.

Example: Chest Measurements

To illustrate the steps in fitting a Gaussian comparison curve, we return to the chest measurement data in Exhibit 9–4. The data are repeated, and the key results of the fitting calculations are shown in Exhibit 9–11. Here, with $N = 5738$, we find the depth of the hinge: $d(H) = [(5738 + 1)/2 + 1]/2 = 1435$. Adding up the n_i from the low end, we find that

$$n_0 + \ldots + n_5 = 707 < 1435 < 1456 = n_0 + \ldots + n_6,$$

so that the lower hinge, H_L, lies between $x_5 = 37.5$ and $x_6 = 38.5$. Interpolation then gives

$$H_L = x_{L-1} + \frac{d(H) - (n_0 + \ldots + n_{L-1}) - 0.5}{n_L} w_L$$

$$= 37.5 + \frac{1435 - 707 - 0.5}{749} 1$$

$$= 38.471.$$

Similarly, summing the n_i from the high end yields

$$n_{10} + \ldots + n_{17} = 1196 < 1435 < 2130 = n_9 + \ldots + n_{17},$$

so that the upper hinge, H_U, lies between $x_8 = 40.5$ and $x_9 = 41.5$. Again, interpolation gives

$$H_U = x_U - \frac{d(H) - (n_{U+1} + \ldots + n_{k+1}) - 0.5}{n_U} w_U$$

$$= 41.5 - \frac{1435 - 1196 - 0.5}{934} 1$$

$$= 41.245.$$

Exhibit 9–11 A Gaussian Comparison Curve for the Chest Measurement Data of Exhibit 9–4, with Double-Root Residuals

i	x_i	n_i	$F((x_i - m)/s)$	$\hat{n}_i$	DRR_i
0		0		0.99	−1.23
	32.5		.00017		
1		3		4.70	−0.71
	33.5		.00099		
2		18		20.57	−0.52
	34.5		.00458		
3		81		71.35	1.13
	35.5		.01701		
4		185		196.22	−0.79
	36.5		.05121		
5		420		427.74	−0.36
	37.5		.12575		
6		749		738.85	0.38
	38.5		.25452		
7		1073		1011.73	1.91
	39.5		.43084		
8		1079		1100.38	−0.64
	40.5		.62261		
9		934		947.28	−0.42
	41.5		.78770		
10		658		647.75	0.41
	42.5		.90059		
11		370		351.01	1.01
	43.5		.96176		
12		92		150.73	−5.34
	44.5		.98803		
13		50		51.31	−0.15
	45.5		.99697		
14		21		13.85	1.76
	46.5		.99938		
15		4		2.96	0.66
	47.5		.99990		
16		1		0.50	0.71
	48.5		.99999		
17		0		0.08	−0.14

From the hinges it is then simple to find

$$m = \tfrac{1}{2}(H_L + H_U) = 39.858 \quad \text{and} \quad s = (H_U - H_L)/1.349 = 2.056.$$

We now use the approximation for $F(z)$ to calculate the fourth column of Exhibit 9–11, the values of $F((x_i - m)/s)$. The differences between adjacent entries, multiplied by $N = 5738$, are the $\hat{n}_i$. The column of double-root residuals, calculated as in Section 9.3, completes the numerical work on this example.

The double-root residuals now tell us how closely the comparison curve follows the data. Immediately, our attention focuses on bin 12, where $DRR = -5.34$. Surely something is amiss in Quetelet's data. The original source of the data, published in 1817, gives the joint frequency distribution of height and chest measurement in each of eleven militia regiments. It has a total count of 5732, and its bin counts differ by as much as 76—in bin 12, it turns out—from the bin counts reported by Quetelet. It seems that Quetelet made some serious copying errors in forming his frequency distribution, but he did not notice the discrepancy that is so evident in Exhibit 9–11.

Except for bin 12, DRR values in Exhibit 9–11 indicate that the fit is reasonable. If we look back at the rootogram in Exhibit 9–9, we may agree that the bar for bin 12 looks a bit low. When we fit the Gaussian comparison curve, however, the double-root residual makes it impossible for this isolated problem to escape notice. We have gained considerably by looking at the fit in this way.

Note also that, because we used the hinges to fit the comparison curve, the one extraordinary bin—which involved a change of only about 1% of the cases—did not have an undue influence on the fit. Correcting the error would change the comparison curve only slightly and thus would not alter the fit at the other bins.

9.5   Suspended Rootograms

In the preceding sections, we concentrated on fitting a comparison curve to our data and on finding the proper residuals. Our approach was different from the approaches we used for other data structures such as y-versus-x and two-way tables, because we fitted the comparison curve to the raw data but calculated residuals in the square-root scale. However, Tukey (1971) describes a way of fitting the comparison curve directly to the rootogram. We now bring the fitted curve and the residuals together in a graphical display.

We recall that Exhibit 9–9, the rootogram for the chest measurement data, tempted us to sketch in a comparison curve. Now that we have fitted a Gaussian comparison curve to that set of data, we can superimpose the fitted curve (actually, its square root, point by point) on the rootogram to produce Exhibit 9–12. Superimposing the fitted curve is a common practice with histograms, but the resulting display does nothing to help us see the residuals, as we should.

We can identify simple "rootogram residuals" in Exhibit 9–12. They are the difference between the height of each bar and the height of the curve at roughly the center of the bin. It is difficult to grasp the whole set of these residuals, however, because we must look along the curve. We can make the differences easier to see by forming the residuals: Writing

$$residual = data - fit$$

is equivalent to putting the comparison curve below the horizontal axis and standing each bar of the rootogram on the curve, near the center of the bin.

suspended rootogram   The resulting display, called a ***suspended rootogram,*** appears in Exhibit 9–13. In Exhibit 9–12 the bars stand on the horizontal axis, and we have to compare them to the curve to see residuals. Now the bars stand on the curve, and the residuals are easily seen as bar-like deviations from the horizontal

Exhibit 9–12   Rootogram for the Chest Measurement Data, with Gaussian Comparison Curve as Background

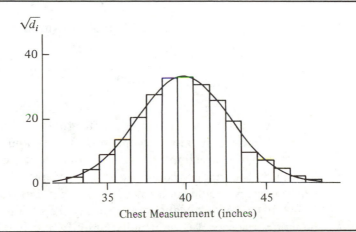

Exhibit 9–13   Rootogram for Chest Measurement Data, Suspended on Gaussian Comparison Curve

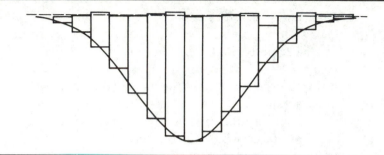

axis. Because a horizontal straight line is a very convenient standard of comparison, we can easily spot large residuals and begin to look for patterns.

To examine the calculations in more detail, we recall (from Section 9.1) that $d_i = n_i/w_i$ and thus the height of the rootogram bar in bin i is $\sqrt{d_i}$. Analogously, we use the fitted count, $\hat{n}_i$, to define $\hat{d}_i = \hat{n}_i/w_i$ so that the rootogram residual in bin i is

$$\sqrt{d_i} - \sqrt{\hat{d}_i}.$$

We judge the size of these residuals by converting the rule of thumb that we use for double-root residuals: A *DRR* is "large" if it is less than -2 or greater than $+2$. Because d_i and $\hat{d}_i$ have n_i and $\hat{n}_i$ as their numerators and w_i as their denominator, we begin with

$$DRR_i = \sqrt{2 + 4n_i} - \sqrt{1 + 4\hat{n}_i},$$

neglect the constants 2 and 1, and multiply through by $1/(2\sqrt{w_i})$ to obtain

$$\frac{DRR_i}{2\sqrt{w_i}} \simeq \sqrt{d_i} - \sqrt{\hat{d}_i}.$$

Thus we regard the rootogram residual in bin i (that is, $\sqrt{d_i} - \sqrt{\hat{d}_i}$) as large if it is (roughly) less than $-1/\sqrt{w_i}$ or greater than $+1/\sqrt{w_i}$.

When all the bins (except the left-open and right-open ones) have the same width, w, these limits for rootogram residuals can be shown as horizontal lines on the suspended rootogram at $-1/\sqrt{w}$ and $+1/\sqrt{w}$. Of course, when the widths vary, we could show lines for each bin, but we seldom do.

Because we always want to study the residuals but seldom need to see the comparison curve, we usually simplify a suspended rootogram and show only the bars for the residuals (along with light lines at $\pm 1/\sqrt{w}$ when we have

Exhibit 9–14   Suspended Rootogram, Showing only Rootogram Residuals, for Chest Measurement Data ($1/\sqrt{w_i} = 1$ for all bins)

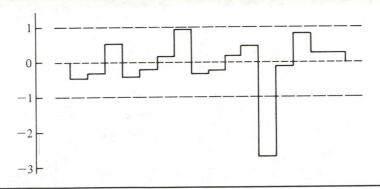

equal-width bins). Exhibit 9–14 illustrates this display. The simplified version is the preferred graphical display for comparing a set of counts and a fitted curve. By showing only the rootogram residuals, Exhibit 9–14 makes better use of plotting space than does Exhibit 9–13, and it is far more effective than a histogram with a superimposed curve.

9.6   Rootograms from the Computer

A general-purpose display for counts in bins could well include several types of information: (1) the bin boundaries, (2) the observed count in each bin, (3) the fitted count in each bin, (4) the ordinary residual ($n_i - \hat{n}_i$), (5) the double-root residual (DRR), (6) the rootogram residual ($\sqrt{d_i} - \sqrt{\hat{d}_i}$), and (7) a suspended rootogram. The constraint of being able to use simple computer terminals, however, forces some compromises. The programs for this chapter display five components:

- the bin number, i
- the observed count, n_i
- the ordinary residual, $n_i - \hat{n}_i$
- the double-root residual, DRR_i, and
- a suspended-rootogram display of the DRR,

as Exhibit 9–15 shows for the chest measurement data. From the observed

Exhibit 9–15   Rootogram Display (based on a Gaussian comparison curve) for the Chest
Measurement Data

BIN	COUNT	RAWRES	DRRES	SUSPENDED ROOTOGRAM
1	0.0	−1.0	−1.23	. -------- .
2	3.0	−1.7	−0.71	. ---- .
3	18.0	−2.6	−0.52	. --- .
4	81.0	9.6	1.13	. ++++++ .
5	185.0	−11.2	−0.79	. ---- .
6	420.0	−7.7	−0.36	. -- .
7	749.0	10.2	0.38	. ++ .
8	1073.0	61.3	1.91	. +++++++++++ .
9	1079.0	−21.4	−0.64	. ---- .
10	934.0	−13.3	−0.42	. --- .
11	658.0	10.2	0.41	. +++ .
12	370.0	19.0	1.01	. ++++++ .
13	92.0	−58.7	−5.34	*---------------- .
14	50.0	−1.3	−0.15	. − .
15	21.0	7.2	1.76	. +++++++++ .
16	4.0	1.0	0.66	. ++++ .
17	1.0	0.5	0.71	. ++++ .
18	0.0	−0.1	−0.14	. − .

IN DISPLAY, VALUE OF ONE CHARACTER IS .2   OO

count and the ordinary residual (in the column headed RAWRES, for "raw
residual") it is easy to reconstruct the fitted count, $\hat{n}_i$: $\hat{n}_i = n_i - \text{RAWRES}_i$.

In order to accommodate the half-open bin at each end, the
suspended-rootogram display is based on the double-root residuals rather than
the rootogram residuals. It appears as a compact display to the right of the
columns of numerical output and shows a (horizontal) bar for each bin on the
same line as the other information for that bin. The plotting character is the
sign of the double-root residual (DRRES), and each horizontal space has the
fixed value of .2. (This fixed amount of space suffices because the double-root
residuals have a natural scale.) Enough spaces are available to show *DRR*
values from −3 to +3, and any value outside this range is marked with a * at
the tip of its bar. In Exhibit 9–15, bin 13 requires this mark. (We referred to
this bin as bin 12 earlier, when we numbered the bins from 0 to $k+1$. The
programs use I = 1, . . . , $k + 2$.) As an aid to drawing in a vertical axis for the
suspended rootogram, the OO in ROOTOGRAM lies where the line can pass
between the Os and is repeated in the same position below the display.

The programs check the number of spaces between the margins set for
the output line. Sixty-five spaces are required for the full display. If fewer than

65 but at least 30 spaces are available, only the numerical columns are printed.

FORTRAN

Two FORTRAN subroutines, RGCOMP and RGPRNT, handle the computations and output for a rootogram display. RGCOMP, in turn, uses the function GAU, which gives the value of the standard Gaussian cumulative distribution function. Separating the computation from the display makes it easy to use the fitted counts or the double-root residuals in other calculations or displays.

For input, the vector X( ) holds the bin boundaries, and the vector Y( ) holds the bin counts. (Y( ) is REAL rather than INTEGER because some frequency distributions include non-integer counts. The most common reason is that one or more data values fell on a bin boundary and were counted as one half in each of the bins that share the boundary.) As in Section 9.1, $Y(I)$ is the count for the bin whose right-hand boundary is $X(I)$. Now I runs from 1 to L (so that $L = k + 2$ in the notation of Section 9.1), and again $X(L)$ is not used. $Y(1)$ and $Y(L)$ hold counts for the unbounded extreme bins and must be zero whenever the data have no unbounded bins.

To fit a Gaussian comparison curve and calculate the double-root residuals, use the following FORTRAN statement

CALL RGCOMP(X, Y, L, MU, SIGMA, YHAT, DRR, MHAT, SHAT, ERR)

where

X( )	is the vector of bin boundaries—X(L) is unused;
Y( )	is the vector of observed counts;
L	is the number of bins;
MU	allows the user to specify the mean of the fitted Gaussian distribution;
SIGMA	allows the user to specify the standard deviation of the fitted Gaussian distribution (if SIGMA = 0.0, the program ignores the values of MU and SIGMA);
YHAT( )	is returned as the vector of fitted counts;
DRR( )	is returned as the vector of double-root residuals;
MHAT	is returned as the mean of the fitted comparison curve;
SHAT	is returned as the standard deviation of the fitted comparison curve;

ERR   is the error flag, whose values are
    0   normal
  91   too few bins ($L < 3$)
  92   a hinge falls in a half-open bin (so that interpolation is not possible).

Then, to produce the rootogram display using the observed counts, the fitted counts, and double-root residuals just calculated, use the FORTRAN statement

    CALL RGPRNT (Y, L, YHAT, DRR, ERR)

where the parameters are as defined for RGCOMP and

ERR   is the error flag, whose values are
    0   normal
  93   margins too narrow for numerical part of display (< 30 spaces available)
  94   margins wide enough for numerical columns but not for graphical display, so graphical display not printed ($30 \leq$ spaces < 65).

Both of these subroutines assume that the data take the form of a frequency distribution. When it is necessary to construct the frequency distribution from a batch of data, the number of bins and the scaling used by the stem-and-leaf display programs generally provide a good starting point.

BASIC

The BASIC program for suspended rootograms is entered with bin boundaries in the array X( ) and bin counts in the array Y( ). As in Section 9.1, Y(I) is the count for the bin whose right-hand boundary is X(I). I runs from 1 to N (so that $N = k + 2$ in the notation of Section 9.1), and X(N) is not used. Y(1) and Y(N) hold the counts for the unbounded extreme bins and must be zero if the data have no unbounded bins. The defined function FNG(Z) is an approximate Gaussian cumulative distribution function; it returns the probability below Z in a standard Gaussian distribution (when $|Z| \leq 5.5$—see Section 9.4).

The program leaves X( ) and Y( ) unchanged and returns fitted counts in C( ) and double-root residuals in R( ).

* 9.7   More on Double Roots

This section briefly brings together several useful facts about double-root residuals. The theoretical background for double-root residuals comes primarily from work on transformations to stabilize the variance of Poisson data. Bartlett (1936) discussed the use of $\sqrt{x}$ and $\sqrt{x + \frac{1}{2}}$ for counted data generated by a Poisson distribution, and others subsequently investigated modifications of these re-expressions. Generally, if the random variable X follows a Poisson distribution with mean m, the re-expressed variable approximately follows a Gaussian distribution whose mean is a function of m and whose variance is approximately $\frac{1}{4}$. The main points are that (1) the variance after the re-expression depends only slightly on m and (2) the approximation becomes better as m grows larger.

In order to do better for small values of m, Freeman and Tukey (1950) suggested the re-expression

$$\sqrt{x} + \sqrt{x + 1}.$$

As Freeman and Tukey (1949) point out (see also Bishop, Fienberg, and Holland, 1975), the average value of $\sqrt{X} + \sqrt{X + 1}$ is well approximated for Poisson X by

$$\sqrt{4m + 1},$$

and its variance is close to 1. It is customary to substitute the estimated or fitted count, $\hat{m}$, for the (unknown) average value, m. The resulting residuals,

$$\sqrt{x} + \sqrt{x + 1} - \sqrt{4\hat{m} + 1},$$

are known as Freeman-Tukey deviates.

For the observed counts, $x = 1, 2, \ldots$, it is easy to check that $\sqrt{x} + \sqrt{x + 1}$ and $\sqrt{4x + 2}$ are only very slightly different. Thus double-root residuals and Freeman-Tukey deviates are essentially equivalent. (Recall that 1 replaces $\sqrt{4x + 2}$ in the definition of the double-root residual when $x = 0$. This is the main difference between using $\sqrt{x} + \sqrt{x + 1}$ and using $\sqrt{4x + 2} = 2\sqrt{x + \frac{1}{2}}$ without special treatment of zero.) The approximate behavior of the Freeman-Tukey deviate is the basis for treating individual *DRR* values as if they were observations from a standard Gaussian distribution.

For descriptive and diagnostic purposes, we can treat the double-root residuals from a fitted frequency distribution as if they were a Gaussian sample. Naturally, the *DRR* values are not all independent; the sum of fitted cell counts must equal the sum of observed cell counts, and it is usually necessary to estimate some parameters from the data—for example, *m* and *s* for the Gaussian comparison curve—but this lack of independence is seldom a serious problem.

Clearly, double-root residuals tell something about goodness of fit between model and data. The almost universally used measure of goodness of fit is the (Pearson) chi-squared statistic,

$$X^2 = \sum_i \frac{(x_i - \hat{m}_i)^2}{\hat{m}_i} .$$

How might the double-root residuals be related to X^2? Because of the approximately Gaussian behavior of the DRR_i,

$$\sum_i DRR_i^2$$

follows roughly a chi-squared distribution. The usual number of degrees of freedom—that is, the number of d.f. appropriate for X^2—takes into account the dependence among the DRR_i. For example, from Exhibit 9–11, we get $\Sigma DRR^2 = 42.48$; and, because there are 18 bins and 2 estimated parameters (besides the total), we should refer this sum to the χ^2_{15} distribution. When we do this, we are led to reject the hypothesis that the differences between the observed and fitted bin counts are due to chance; in fact, $p < .0005$. The value of X^2 for this same fit is 37.13, which is almost significant at the .001 level. Both measures indicate strongly that the fit is not satisfactory. (Almost all the difference between X^2 and ΣDRR^2 comes from the bin centered at 44 inches; DRR^2 is 28.52, while the contribution to X^2 is 22.88.) The practice of beginning by looking at the individual DRR_i will call early attention to any bin where the fit is poor. Forming ΣDRR^2 as a second step will then provide an overall measure (in case the fit is generally poor but not unusually bad in any one cell).

In using X^2, it is customary to combine bins at either end of the frequency distribution until every bin has a fitted count no smaller than 1. While some further research is required, this restriction does not seem to be necessary for double-root residuals. (Tukey suggests that we can make a rather satisfactory allowance for small fitted counts by subtracting $\Sigma(1 - \hat{n}_i)^2$, where only bins with $\hat{n}_i < 1$ contribute to the sum, from the conventional number of degrees of freedom.)

References

Bartlett, M.S. 1936. "The Square Root Transformation in Analysis of Variance." *Supplement to the Journal of the Royal Statistical Society* 3:68–78.

Bishop, Y.M.M., S.E. Fienberg, and P.W. Holland. 1975. *Discrete Multivariate Analysis: Theory and Practice.* Cambridge, Mass.: MIT Press.

Derenzo, Stephen E. 1977. "Approximations for Hand Calculators Using Small Integer Coefficients." *Mathematics of Computation* 31:214–225.

Freeman, Murray F., and John W. Tukey. 1949. "Transformations Related to the Angular and the Square Root." Memorandum Report 24. Statistical Research Group, Princeton University, Princeton, N.J.

Freeman, Murray F., and John W. Tukey. 1950. "Transformations Related to the Angular and the Square Root." *Annals of Mathematical Statistics* 21:607–611.

Huffman, Sandra L., A.K.M. Alauddin Chowdhury, and W. Henry Mosley. 1979. "Difference Between Postpartum and Nutritional Amenorrhea" (reply to Frisch and McArthur). *Science* 203 (2 March 1979), pp. 922–923.

Pollard, R. 1973. "Collegiate Football Scores and the Negative Binomial Distribution." *Journal of the American Statistical Association* 68:351–352.

Quetelet, A. 1846. *Lettres à S.A.R. le Duc Régnant de Saxe-Cobourg et Gotha, sur la Théorie des Probabilités, Appliquée aux Sciences Morales et Politiques.* Brussels: M. Hayez.

Tukey, John W. 1971. *Exploratory Data Analysis,* limited preliminary edition, vol. III. Reading, Mass.: Addison-Wesley.

Tukey, John W. 1972. "Some Graphic and Semigraphic Displays." In T.A. Bancroft, ed., *Statistical Papers in Honor of George W. Snedecor.* Ames: Iowa State University Press, pp. 293–316.

```
5000 REM SUSPENDED ROOTOGRAM
5010 REM ON ENTRY X() HOLDS BIN BOUNDARIES (N-1 OF THEM).
5020 REM Y() HOLDS BIN COUNTS (N OF THEM),  N=# OF BINS.
5030 REM Y(1) AND Y(N) ARE ASSUMED TO HOLD COUNTS BELOW X(1)
5040 REM AND ABOVE X(N), RESPECTIVELY.  THEY MUST BE
5050 REM SET TO ZERO IF THEY ARE NOT NEEDED.
5060 REM FNG(X) IS ASSUMED TO BE DEFINED AS THE CUMULATIVE
5070 REM PROBABILITY FUNCTION TO BE FIT (THE GAUSSIAN BY DEFAULT).
5080 REM IF V1<0 REQUESTS VALUES FOR MEAN AND STANDARD DEVIATION
5090 REM AND SKIPS FITTING PROCEDURE.
5100 REM ON EXIT, X() AND Y() ARE UNCHANGED. THE FITTED COUNTS
5110 REM ARE IN C(), AND THE DOUBLE-ROOT RESIDUALS ARE IN R().
5120 REM

5130 IF N >= 3 THEN 5160
5140 LET E9 = 91
5150 RETURN

5160 REM FIND TOTAL COUNT

5170 LET A = 0
5180 FOR I = 1 TO N
5190    LET A = A + Y(I)
5200 NEXT I
5210 IF V1 > 0 THEN 5290

5220 REM GET USER-SUPPLIED PARAMTERS

5230 PRINT TAB(M0);"MEAN, STANDARD DEVIATION";
5240 INPUT L1,S1
5250 IF S1 > 0 THEN 5590
5260 PRINT TAB(M0);"S.D. MUST BE > 0, RE-ENTER ";
5270 GO TO 5230

5280 REM FIND HINGES

5290 LET A1 = ( INT((A + 1) / 2) + 1) / 2
5300 IF A1 > Y(1) THEN 5330
5310 PRINT TAB(M0);"HINGE IN LEFT-OPEN BIN  IN ROOTOGRAM"
5320 STOP
5330 LET A2 = Y(1)
5340 FOR I = 2 TO N - 1
5350    LET A2 = A2 + Y(I)
5360    IF A2 >= A1 THEN 5400
5370 NEXT I
5380 PRINT TAB(M0);"HINGE IN RIGHT-OPEN BIN  IN ROOTOGRAM"
5390 STOP
```

```
5400 REM FIND LOW HINGE BY INTERPOLATION AND PUT IN L2

5410 LET A2 = A2 - Y(I)

5420 LET L2 = X(I - 1) + (X(I) - X(I - 1)) * (A1 - A2 - .5) / Y(I)

5430 REM NOW FIND THE HIGH HINGE

5440 IF A1 <= Y(N) THEN 5380
5450 LET A4 = Y(N)
5460 FOR I = N - 1 TO 2 STEP - 1
5470    LET A4 = A4 + Y(I)
5480     IF A4 >= A1 THEN 5510
5490 NEXT I
5500 GO TO 5310
5510 LET A4 = A4 - Y(I)
5520 LET L3 = X(I) - (X(I) - X(I - 1)) * (A1 - A4 - .5) / Y(I)

5530 REM L2 AND L3 ARE NOW THE HINGES.  USE THE MIDHINGE AS A CENTER
5540 REM AND HINGESPREAD/1.349 AS A SCALE IN GAUSSIAN.

5550 LET L1 = (L2 + L3) / 2
5560 LET S1 = (L3 - L2) / 1.349

5570 REM C7 ACCUMULATES CUMULATIVE PROBABILITY
5580 REM IS TOTAL COUNT. C() GETS FITTED COUNT,
5590 REM R() GETS DOUBLE-ROOT RESIDUALS.

5600 LET C7 = 0
5610 FOR I = 1 TO N - 1
5620    LET C8 = FNG((X(I) - L1) / S1)
5630    LET C(I) = A * (C8 - C7)
5640    LET R(I) = SQR(2 + 4 * Y(I)) - SQR(1 + 4 * C(I))
5650    IF Y(I) > 0 THEN 5670
5660    LET R(I) = 1 - SQR(1 + 4 * C(I))
5670    LET C7 = C8
5680 NEXT I

5690 REM NOW HANDLE RIGHT-OPEN BIN

5700 LET C(N) = A * (1 - C7)
5710 LET R(N) = SQR(2 + 4 * Y(N)) - SQR(1 + 4 * C(N))
5720 IF Y(N) > 0 THEN 5740
5730 LET R(N) = 1 - SQR(1 + 4 * C(N))

5740 REM
5750 REM PRINT ROOTOGRAM RESULTS
5760 REM

5770 LET M1 = M9 - M0 + 1
5780 IF M1 > 30 THEN 5810
5790 PRINT TAB(M0);"PAGE TOO NARROW TO DISPLAY ROOTOGRAM RESULTS"
5800 RETURN
```

```
5810 REM SET UP TABS

5820 LET T1 = M0 + 4
5830 LET T2 = T1 + 8
5840 LET T3 = T2 + 8
5850 LET T4 = T3 + 8

5860 REM R3 IS PRINTING FLAG: 0= PRINT TABLE,
5870 REM 1 = PRINT TABLE AND ROOTOGRAM, 2 = ROOTOGRAM ONLY

5880 LET R3 = 1
5890 IF M1 >= 60 THEN 5910
5900 LET R3 = 0
5910 PRINT
5920 PRINT TAB(M0);"BIN#"; TAB(T1 + 1);"COUNT"; TAB(T2);"RAW RES";
5930 PRINT TAB(T3);"D-R RES";
5940 IF R3 = 0 THEN 5970

5950 REM HEADING FOR ROOTOGRAM DISPLAY

5960 PRINT TAB(T4 + 4);"SUSPENDED ROOTOGRAM";
5970 PRINT
5980 PRINT
5990 FOR I = 1 TO N
6000    LET R1 = Y(I) - C(I)
6010    LET R0 = 1
6020    PRINT TAB(M0);I;
6030    IF R3 = 2 THEN 6080
6040    PRINT TAB(T1); FNR(Y(I)); TAB(T2); FNR(R1);
6050    LET R0 = 2
6060    PRINT TAB(T3); FNR(R(I));
6070    IF R3 = 0 THEN 6420

6080    REM PUT ONE LINE OF ROOTOGRAM IN P()

6090    LET O1 = ASC(" ")
6100    FOR J = 1 TO 32
6110       LET P(J) = O1
6120    NEXT J
6130    LET P(6) = ASC(".")
6140    LET P(27) = ASC(".")
6150    LET J1 = 0
6160    IF R(I) = 0 THEN 6360
6170    LET X1 = FNC(5 * ABS(R(I)))
6180    IF X1 <= 15 THEN 6200
6190    LET X1 = 15
6200    IF R(I) > 0 THEN 6290
```

```
6210    REM CONSTRUCT ROOTOGRAM LINE FOR RESIDUAL < 0

6220    LET Jl = 16

6230    FOR J = Jl TO Jl - Xl STEP - l
6240      LET P(J) = ASC("-")
6250    NEXT J
6260    IF Xl < 15 THEN 6360
6270    LET P(1) = ASC("*")
6280    GO TO 6360

6290    REM CONSTRUCT ROOTOGRAM LINE FOR RESIDUAL>0

6300    LET Jl = 17 + Xl
6310    FOR J = 17 TO Jl
6320      LET P(J) = ASC("+")
6330    NEXT J
6340    IF Xl < 15 THEN 6360
6350    LET P(32) = ASC("*")
6360    IF Jl >= 27 THEN 6380
6370    LET Jl = 27
6380    PRINT TAB(T4);
6390    FOR J = 1 TO Jl
6400      PRINT CHR$(P(J));
6410    NEXT J
6420    PRINT
6430 NEXT I

6440 REM GO BACK TO PRINT ROOTOGRAM?

6450 IF R3 >= 1 THEN 6510
6460 LET R3 = 2
6470 LET T4 = M0 + 4
6480 IF M1 > T4 + 30 THEN 5950
6490 PRINT TAB(M0);"PAGE TOO NARROW FOR ROOTOGRAM"
6500 GO TO 6550
6510 PRINT

6520 REM WRAPUP

6530 PRINT TAB(T4 + 15);"/"; CHR$(92)
6540 PRINT TAB(M0);"IN DISPAY, VALUE OF ONE SPACE IS .2"
6550 RETURN
```

```
      SUBROUTINE RGCOMP(X, Y, L, MU, SIGMA, YHAT, DRR, MHAT, SHAT, ERR)
C
      INTEGER L, ERR
      REAL X(L), Y(L), YHAT(L), DRR(L), MU, SIGMA, MHAT, SHAT
C
C PERFORM THE COMPUTATIONS FOR A SUSPENDED ROOTOGRAM.
C X(1), ..., X(L) ARE THE BIN BOUNDARIES, AND Y(1),
C ..., Y(L) ARE THE BIN COUNTS (I.E., CELL FREQUENCIES).
C THE COUNT  Y(I)  CORRESPONDS TO THE BIN WHOSE RIGHT
C BOUNDARY IS  X(I) .  THE  BIN WHOSE RIGHT
C BOUNDARY IS  X(1)  IS OPEN TO THE LEFT.  ALSO
C X(L) IS NOT USED, SO THAT Y(L) COUNTS ALL DATA VALUES
C TO THE RIGHT OF X(L-1).
C A GAUSSIAN COMPARISON CURVE IS USED, AND ITS CENTER
C AND SCALE ARE DETERMINED BY THE HINGES OF THE DATA
C (FOUND BY LINEAR INTERPOLATION).
C IF  SIGMA  IS NOT EQUAL TO ZERO, THEN THE FITTING PROCESS IS SKIPPED
C AND THE VALUES OF  MU  AND  SIGMA  PASSED IN ARE USED FOR THE
C COMPARISON CURVE.  IF  SIGMA  IS EQUAL TO ZERO, THE VAUES OF BOTH
C MU  AND  SIGMA ARE IGNORED.
C ON EXIT,  MHAT  CONTAINS THE FITTED MEAN OF THE GAUSSIAN
C COMPARISON CURVE, AND  SHAT  CONTAINS THE FITTED STANDARD
C DEVIATION,  YHAT()  CONTAINS THE  L  FITTED COUNTS, AND
C DRR()  CONTAINS THE DOUBLE-ROOT RESIDUALS.
C
C LOCAL VARIABLES
C
      INTEGER I, K, LP1, LP1MI
      REAL  D, HL, HU, P, PL, T, TN, YH
C
      IF(L .GE. 3) GO TO 5
      ERR = 91
      GO TO 999
    5 K = L - 1
      TN = 0.0
      DO 10 I = 1, L
        TN = TN + Y(I)
   10 CONTINUE
C
C IF  MU  AND  SIGMA  WERE SPECIFIED, DONT BOTHER TO FIT THEM FROM THE
C DATA.  CUE IS  NON-ZERO  SIGMA.
C
      IF(SIGMA .GT. 0.0) GO TO 80
C
      D = 0.5 * (1.0 + AINT(0.5 * (TN + 1.0)))
```

```
C
C   IF LOWER HINGE FALLS IN LEFT-OPEN BIN, ERROR.
C
        IF(D .GT. Y(1)) GO TO 20
          ERR = 92
          GO TO 999
   20 T = Y(1)
        DO 30 I = 2, K
          T = T + Y(I)
          IF(T .GE. D) GO TO 40
   30 CONTINUE
C
C   LOWER HINGE FALLS IN RIGHT-OPEN BIN -- ERROR.
C
        ERR = 92
        GO TO 999
C
C   FIND LOWER HINGE BY INTERPOLATION.
C
   40 T = T - Y(I)
        HL = X(I-1) + (X(I) - X(I-1)) * (D - T - 0.5) / Y(I)
C
C   NOW PERFORM SIMILAR CHECKS AND FIND UPPER HINGE.
C
        IF(D .GT. Y(L)) GO TO 50
          ERR = 92
          GO TO 999
   50 T = Y(L)
        LP1 = L + 1
        DO 60 I = 2, K
          LP1MI = LP1 - I
          T = T + Y(LP1MI)
          IF(T .GE. D) GO TO 70
   60 CONTINUE
C
        ERR = 92
        GO TO 999
C
   70 T = T - Y(LP1MI)
        HU = X(LP1MI) - (X(LP1MI) - X(LP1MI-1)) * (D - T - 0.5) /
     1    Y(LP1MI)
C
C   USE  MHAT = MID-HINGE  FOR CENTERING AND  SHAT =
C   (H-SPREAD)/1.349  FOR SCALE.  (SHAT  IS AN ESTIMATE OF THE
C   STANDARD DEVIATION FOR THE FITTED GAUSSIAN
C   COMPARISON CURVE.)
C
        MHAT = (HL + HU) / 2.0
        SHAT = (HU - HL) / 1.349
C
        GO TO 90
```

```
C
C   SKIP TO HERE IF  MU  AND  SIGMA  WERE SPECIFIED
C
   80 MHAT = MU
      SHAT = SIGMA
C
   90 PL = 0.0
      DO 100 I = 1, K
          NOTE: SOME FORTRANS MAY WANT THE ARGUMENT OF GAU() TO
              BE A TEMPORARY REAL SCALAR.
C
          P = GAU((X(I) - MHAT) / SHAT)
          YH = TN * (P - PL)
          YHAT(I) = YH
          DRR(I) = SQRT(2.0 + 4.0 * Y(I)) - SQRT(1.0 + 4.0 * YH)
          IF(Y(I) .EQ. 0.0) DRR(I) = 1.0 - SQRT(1.0 + 4.0 * YH)
          PL = P
  100 CONTINUE
      YH = TN * (1.0 - PL)
      YHAT(L) = YH
      DRR(L) = SQRT(2.0 + 4.0 * Y(L)) - SQRT(1.0 + 4.0 * YH)
      IF(Y(L) .EQ. 0.0) DRR(L) = 1.0 - SQRT(1.0 + 4.0 * YH)
  999 RETURN
      END




      SUBROUTINE RGPRNT(Y, L, YHAT, DRR, ERR)
C
      INTEGER L, ERR
      REAL Y(L), YHAT(L), DRR(L)
C
C  PRINT, BIN BY BIN, THE OBSERVED COUNT, THE RAW
C  RESIDUAL, THE DOUBLE-ROOT RESIDUAL, AND AN ABBREVIATED
C  DISPLAY OF THE DOUBLE-ROOT RESIDUAL.
C  Y(1), ..., Y(L)  ARE THE BIN COUNTS.
C  YHAT  CONTAINS THE FITTED COUNTS, AND
C  DRR  CONTAINS THE DOUBLE-ROOT RESIDUALS.
C
C  LOCAL VARIABLES
C
      INTEGER BL, BO, DOT, I, J, MIN, NBL, NMIN, NPL, PL, STAR
      REAL RES
C
C  FUNCTION
      INTEGER FLOOR
C
      COMMON /CHRBUF/ P, PMAX, PMIN, OUTPTR, MAXPTR, OUNIT
      INTEGER P(130), PMAX, PMIN, OUTPTR, MAXPTR, OUNIT
C
      DATA BL, DOT, MIN, PL, STAR /1H , 1H., 1H-, 1H+, 1H*/
```

```
C
C   IS PRINT LINE WIDE ENOUGH TO HOLD THE COLUMNS OF NUMBERS.
C
      IF(PMAX .GE. 30) GO TO 10
         ERR = 93
         GO TO 999
C
C   PRINT LINE MAY BE ADEQUATE FOR NUMBERS BUT NOT DISPLAY.
C
   10 IF(PMAX .GE. 65) GO TO 30
      ERR = 94
C
C   PRINT ONLY THE OBSERVED COUNTS AND THE TWO TYPES OF
C   RESIDUALS.
C
      WRITE(OUNIT, 5010)
 5010 FORMAT(1X,3HBIN,3X,5HCOUNT,3X,6HRAWRES,3X,5HDRRES/)
C
      DO 20 I = 1, L
         RES = Y(I) - YHAT(I)
         WRITE(OUNIT, 5020) I, Y(I), RES, DRR(I)
   20 CONTINUE
 5020 FORMAT(1X,I3,2X,F6.1,4X,F5.1,4X,F5.2)
C
      GO TO 999
C
C   PRINT THE TABLE AND THE DISPLAY.
C
   30 WRITE(OUNIT, 5030)
 5030 FORMAT(1X,3HBIN,3X,5HCOUNT,3X,6HRAWRES,4X,5HDRRES,
     1   7X,19HSUSPENDED ROOTOGRAM/)
C
      DO 120 I = 1, L
         RES = Y(I) - YHAT(I)
         IF(DRR(I) .NE. 0.0) GO TO 40
         WRITE(OUNIT, 5040) I, Y(I), RES, DRR(I)
 5040    FORMAT(1X,I3,2X,F6.1,4X,F5.1,4X,F5.2,8X,1H.,20X,1H.)
         GO TO 120
   40    IF(DRR(I) .GT. 0.0) GO TO 80
C
C   HANDLE LINES WITH NEGATIVE DRR.
C   THERE ARE FOUR CASES:
C     -S END IN * TO INDICATE OVERFLOW,
C     -S OVERWRITE DOT BUT FIT ON LINE,
C     NO BLANKS BETWEEN DOT AND  -S,   AND
C     AT LEAST ONE BLANK BETWEEN DOT AND -S.
C
         NMIN = - FLCOR(5.0 * DRR(I))
         IF(NMIN .GT. 10) GO TO 60
         IF(NMIN .LT. 10) GO TO 50
         WRITE(OUNIT, 5050) I, Y(I), RES, DRR(I),
     1      (BL,J=1,5), DOT, (MIN, J=1,10), (BL, J=1,10), DOT
 5050    FORMAT(1X,I3,2X,F6.1,4X,F5.1,4X,F5.2,3X,32A1)
         GO TO 120
```

```
C
   50    NBL = 10 - NMIN
         WRITE(OUNIT, 5050) I, Y(I), RES, DRR(I),
    1      (BL, J=1,5), DOT, (BL, J=1,NBL), (MIN, J=1, NMIN),
    2      (BL, J=1,10), DOT
         GO TO 120
C
   60    BO = BL
         IF(NMIN .LE. 15) GO TO 70
         NMIN = 15
         BO = STAR
   70    NBL = 16 - NMIN
         WRITE(OUNIT, 5050) I, Y(I), RES, DRR(I),
    1      (BO,J=1,NBL), (MIN,J=1,NMIN), (BL,J=1,10), DOT
         GO TO 120
C
C HANDLE LINES WITH POSITIVE  DRR.
C THERE ARE FOUR CASES:
C   +S END IN * TO INDICATE OVERFLOW,
C   +S OVERWRITE DOT BUT FIT ON LINE,
C   NO BLANKS BETWEEN DOT AND +S, AND
C   AT LEAST 1 BLANK BETWEEN DOT AND +S
C
   80    NPL = - FLOOR(-5.0 * DRR(I))
         IF(NPL .GT. 10) GO TO 100
         IF(NPL .LT. 10) GO TO 90
         WRITE(OUNIT, 5050) I, Y(I), RES, DRR(I),
    1      (BL, J=1,5), DOT, (BL, J=1,10), (PL, J=1,10), DOT
         GO TO 120
C
   90    NBL = 10 - NPL
         WRITE(OUNIT, 5050) I, Y(I), RES, DRR(I),
    1      (BL,J=1,5), DOT, (BL,J=1,10), (PL,J=1,NPL), (BL,J=1,NBL), DOT
         GO TO 120
C
  100    BO = BL
         IF(NPL .LE. 15) GO TO 110
         NPL = 15
         BO = STAR
  110    NBL = 16 - NPL
         WRITE(OUNIT, 5050) I, Y(I), RES, DRR(I),
    1      (BL,J=1,5), DOT, (BL,J=1,10), (PL,J=1,NPL), (BO,J=1,NBL)
C
  120 CONTINUE
C
      WRITE(OUNIT, 5060)
 5060 FORMAT(/1X,40HIN DISPLAY, VALUE OF ONE CHARACTEP IS .2,
    1      7X,2HOO/)
C
  999 RETURN
      END
```

Appendix A

Computer Graphics

Many exploratory techniques are graphical or have a graphical component. Computer programs to produce displays must be able to accommodate widely disparate batches of data and adjust the display parameters to show each batch clearly. Decisions about display formats reflect the purposes of the programs. Displays for exploratory data analysis often do best when formatting decisions are different from the decisions common to traditional practice in computer graphics. This appendix discusses the philosophy, the default display-formatting algorithms, and the technical details of the display programs in this book.

A.1 Terminology

The vocabulary of computer graphics has developed from work in several disciplines and is not standardized. This section defines one common terminology for use in this appendix.

page

A graph or display of data is a representation of data values on some surface—typically paper or the screen of a cathode ray tube (CRT). In this appendix this surface is called the *page* regardless of its true physical form. The type of display determines how the data structure and data values are translated into spatial relationships and symbolic representation.

data coordinates data space

Most rudimentary graphs convey information only through the spatial relationships of points on the page. Conceptually, these points have *data coordinates* in a *data space* determined by the numeric values of the data items or by their place in a data structure—for example, row or column number or group identity. To construct a graph, data coordinates must be mapped on the page into positions described in physical *plotter coordinates*. In printer plotting, these plotter coordinates consist of a line specification and a character position in that line. Data coordinates are translated into plotter coordinates by using a *scale factor* for each coordinate dimension and by pairing at least one data-space point—typically the plot origin, a corner of the plot, or the margins—with a plotter-space position. For example, a simple *x-y* plot might specify the upper-left character position on the page to be data coordinates (0,100), each horizontal one-character print space as 5 *x*-units (*x*-scale), and each vertical line space as 10 *y*-units (*y*-scale). A multiple boxplot might specify the left margin of the page as *x*-value −50, each horizontal one-character space as 2 *x*-units, and each 3 lines as a group identity.

plotter coordinates

scale factor

semigraphic

Exploratory displays are often *semigraphic* (Tukey, 1972)—that is, they choose printed characters to augment the information conveyed by their position on the page. When the character printed is selected from a set of equally-spaced codes—for example, digits—an additional scale factor is needed to map this spacing into data coordinates. At other times the character can symbolize the nature of a data value—for example, that it is the median—or an aspect of its identity—for example, which of five groups it belongs to.

viewport

A display is realized on some region of the page. The plotter coordinates of the edges of this region define the *viewport*. The programs in this book use special symbols at the edges of the display to indicate data points that have been mapped into plotter coordinates outside the viewport. In some displays a corresponding decision is made to exclude or treat specially data values beyond some *data bounds*. (The limits on displays are often called the "plot window," but the subtle difference between the data-space window (data bounds) and the plotter-space window (viewport) can be lost.) For example, a request for a 15-line condensed plot (see Chapter 4) is a viewport specification. Deciding to display only the *y*-values between 0 and 50 is a data-bound specification. Either or both could be valuable in tailoring a condensed plot to a specific need.

data bounds

A.2   Exploratory Displays

Displays for exploratory data analysis should be modified for computer generation in ways that reflect their use. The programs in this book follow several rules to achieve effective exploratory displays:

1. Displays should be structured so that features of the data can be seen easily.
2. Display scales and formats should be resistant to the effects of extraordinary points but should clearly indicate such points when they are present.
3. Displays must be concise so that several can be produced on an interactive computer terminal without lengthy delays. (30 seconds per plot at 30 characters per second is a reasonable maximum.)

Other common requirements of computer displays are less important in exploratory work and have been sacrificed when necessary. For example, the three rules just given contradict the common rule that every data point must be displayed. Exploratory displays often exclude extraordinary points from the main part of the display so that patterns in the main body of the data stand out. (The programs in this book always allow the data analyst to override this decision.) Features such as extensive axis labels and sophisticated options for display titles are desirable in ordinary computer graphics but are unnecessary here and have not been included in these programs. (Nevertheless, these features can be valuable if they are designed to be concise. Implementors of these programs—and especially implementors adding them to an existing high-level program— should consider adding these features.)

A.3   Resistant Scaling

A single plot-scaling algorithm serves all of the programs for displays in this book. It uses the H-spread (see Chapter 2) as an estimate of the variability of a batch. We first define a

$$\text{step} = 1.5 \times \text{H-spread}.$$

The (inner) fences are then placed at one step beyond each hinge. (Chapter 3

adjacent value

discusses fences from a data analysis perspective.) The outermost data value on each end that is not beyond the inner fence is called an *adjacent value*. The high and low adjacent values provide a good frame for the body of a data batch. Data values beyond the fences are treated as outliers. Data values between the fences are displayed in ways that make their important features clearly visible.

For easy comprehension, displays should be made in simple units. Thus, for printer plots, the data-space size of one line or one character space should be easy to understand and easy to count. We call numbers suitable for this purpose *nice numbers*. Nice numbers have the form $m \times 10^e$, where e is an integer and m is selected from a restricted set of numbers. These programs select between two sets of numbers for m: $\{1, 2, 5, 10\}$ and $\{1, 1.5, 2, 2.5, 3, 4, 5, 7, 10\}$. The 1 is redundant, but including both 1 and 10 simplifies the programs. These sets of numbers are chosen to be approximately equally spaced in their logarithms while still being integers or half-integers. This spacing limits the error introduced in approximating a number by a nice number: In the first set, the approximation error is no more than about 40% of the number; in the second set, it is no more than about 20% of the number. (This error bound can be cut to 18% by including 1.25.)

nice numbers

nice position width

Display scaling is accomplished by finding a *nice position width* for each dimension of the display. This is the largest data-space width of a plot position (character space or line), chosen from a set of nice number choices, such that the available number of plot positions (viewport) will accommodate all the numbers between the data bounds. Some displays can have lines labeled -0, which appear to the display scaling algorithm as extra plot positions. The programs allow for these extra lines when they are needed. Because the width of a plot position is approximated with a greater or equal nice value, the range actually covered by a display will generally be slightly larger than that indicated by the data bounds.

A.4   Printer Plots

All the displays in this book have been designed or modified to be produced on a typewriter-style device and are intended to be used interactively. Both of these constraints influence the design of display formatting decisions.

Each display is produced line by line, starting from the top of the page. This may be different from the way the display would be drawn by hand or on a more sophisticated graphics device. All displays start at or near the left margin, so little time is wasted on an interactive terminal spacing out to the display. Axis labels are placed on the left to keep empty lines short.

Printer plots are inherently granular. Rounding each data-space coordinate to one of, say, 50 character positions distorts a display, although this rarely diminishes the display's usefulness in an exploratory analysis. Printed characters are usually taller than they are wide. As a result, the horizontal and vertical scales of a plot may not be comparable; and, for example, the apparent slope of a line may not be closely related to the actual slope value. The displays in this book openly treat each axis differently.

These inconveniences are balanced by a wide choice of plotting characters. Almost every display in this book takes advantage of this choice either to report numbers with greater precision (stem-and-leaf display, condensed plot) or to code important characteristics (boxplot, coded table). The programming languages used here have dictated some choices of codes (see Appendix C). Other choices would be reasonable in other languages.

The experienced programmer reading these programs will probably find FORTRAN especially stifling in this respect. The FORTRAN language has a very restricted character set and limited abilities in character manipulation. Occasionally, our attempts to write clear, portable, easily understood programs for graphics may have been stymied by the FORTRAN language, and for this we ask the programming reader's indulgence.

A.5   Display Details

Each of the exploratory displays considered in the text implements different aspects of the methods described in this appendix. This section discusses each display specifically. The discussion assumes knowledge of the displays themselves.

Stem-and-leaf displays (Chapter 1) bound the data strictly at the adjacent values. Data values beyond these bounds appear on special HI and LO lines (even if they might have fit as the most extreme numbers on the final stems) and do not affect the scale. The display scale is the smallest nice number (with m chosen from the set $\{1, 2, 5, 10\}$) such that no more than $10 \times \log_{10}n$ lines are needed to display all data values between the fences. When both positive and negative numbers are in the batch, room is allocated for the -0 stem. The selection of m determines the form of the display. Regardless of the display scale, the character codes always have the same scale, 1×10^e, and thus hold the next digit of each number after the last digit forming the stem. The horizontal viewport is the line length specified by the line margins. Lines overflowing the right margin end with a $*$ to indicate the omission of points falling beyond the viewport.

Boxplots (Chapter 3) do not bound the data at all because one of the primary purposes of a boxplot is to display outliers. One horizontal character position is scaled to the smallest nice number (using $m \ \varepsilon \ \{1, 1.5, 2, 2.5, 3, 4, 5, 7, 10\}$) that accommodates the range of the data on the available line width. Special codes are assigned to outliers, to the median, and to the hinges, as detailed in Chapter 3. When multiple boxplots are generated, one or three lines are allocated to each group depending upon the form of the boxplot.

Condensed plots (Chapter 4) bound data in both dimensions implicitly through the plot scaling. Scales in x and y are nice numbers with $m \ \varepsilon \ \{1, 1.5, 2, 2.5, 3, 4, 5, 7, 10\}$. The y-scale allows for a -0 line if positive and negative y-values are present. The scales are the smallest nice numbers that accommodate the data between the adjacent values in each dimension within the specified number of lines (y-scale) or allowed line width (x-scale). As a result, fewer lines than the specified maximum may be needed, and data values beyond the fences may fit within the viewport. Data values mapped outside the viewport are indicated with special characters at the edges of the plot, as described in Section 4.6. Character codes are scaled according to C, the number of characters specified for the display. The vertical line size in data-space units is divided into C equal intervals. Plot symbols starting with 0 and counting through successive integers are assigned outward from the edge of the interval nearer to zero. Options allow the user to specify data bounds to supplant the adjacent values in scale calculation, viewport (as number of lines—the x-dimension viewport is defined by the line width), and number of codes (as number of characters). These options allow the display to focus on any segment of the data, enlarge it to any size, and magnify the vertical precision up to 10 times by coding. The default settings of these options are designed to produce the display most likely to be useful in an exploratory analysis.

Coded tables (Chapter 7) require no scaling. The data structure determines the format on the page. One line is allocated per row of the table. Two character positions are allocated per column of the table. Codes are scaled to identify data-value characteristics with respect to the data batch based upon the hinges and fences as detailed in Section 7.1.

Suspended rootograms (Chapter 9) need no special display scaling; the numbers displayed are automatically well-scaled. One line is allocated to each bin and contains both numeric and graphical output. The rootogram display plots the double-root residuals, which, as computed, are expected to behave as if drawn from a standard Gaussian distribution. The display allocates one character position to a unit of .2 in data (double-root-residual) space.

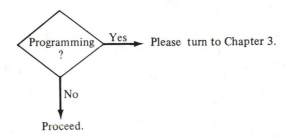

Appendix B

Utility Programs

B.1 BASIC

```
10 REM UTILITY PROGRAMS USED BY THE EDA PROGRAMS.
20 REM ALL VARIABLES ARE GLOBAL.   UTILITY FUNCTION DEFINITIONS
30 REM COME FIRST (AS REQUIRED BY SOME BASIC IMPLEMENTATIONS).
40 REM CONVENTIONS: X(),Y() -- DATA ARRAYS OF LENGTH N.  W() -- WORK
50 REM ARRAY. R() AND C() HOLD ROW AND COLUMN SUBSCRIPTS WHEN Y()
60 REM HOLDS A MATRIX, UTILITY SPACE OTHERWISE. P()--PRINT ARRAY.
70 REM 1000 SORT W()1500 NICE NUMBER 3000 SORT X TO W 3800 SWAP Y&W.
80 REM 1200 SORT X WITH Y 1900 NICE POSN WIDTH 3300 COPY Y TO W & SORT
90 REM 1400 SWAP X&Y 2500 INFO ON W() 3600 COPY Y TO W SELECTIVELY
100 REM INITIALIZER
110 REM FUNCTION DEFINITIONS--THESE ARE USED IN VARIOUS SUBROUTINES
120 REM
130 REM NICE INTEGER PART FUNCTION--ROUNDS TOWARDS ZERO

140 DEF FNI(X) = INT((1 + E0) * ABS(X)) * SGN(X)

150 REM NICE FLOOR FUNCTION--ROUNDS DOWN: NOTE BASIC INT(X) IS A FLOOR
160 REM FUNCTION.  IF IT ISN'T, FIX IT HERE.

170 DEF FNF(X) = INT(X + E0)
```

301

```
180 REM BASE 10 LOG IN CASE NOT A SYSTEM FUNCTION
190 REM NOTE: LOG10(X)=LOG(X)/LOG(10)
200 REM NOTE PROTECTION FROM X<=0 BY ADDING E0 AND ABS

210 DEF FNL(X) = LOG( ABS(X) + (1 - ABS( SGN(X))) * E0) / LOG(10)

220 REM FUNCTION TO SELECT THE HIGH ORDER T8 DIGITS OF X

230 DEF FNT(X) = FNF(X / FNU(X)) * FNU(X)

240 REM CLEAN POWER OF 10 FOR TRUNCATING

250 DEF FNU(X) = 10 ^ ( FNF( FNL(X)) - T8 + 1)

260 REM ROUNDING FUNCTION. ROUND TO R0 PLACES FROM DECIMAL POINT.

270 DEF FNR(X) = FNI( ABS(X) * 10 ^ R0 + .5) / 10 ^ R0 * SGN(X)

280 REM RETRIEVES THE X-TH ELEMENT OF W(),
290 REM AVERAGING IF X ISN'T AN INTEGER.

300 DEF FNM(X) = (W( INT(X)) + W( INT(X + .5))) / 2

310 REM RETRIEVE THE Y-TH ELEMENT OF X() JUST LIKE FNM

320 DEF FNN(Y) = (X( INT(Y)) + X( INT(Y + .5))) / 2

330 REM POSITION FUNCTION FOR PLOTTING.
340 REM CALLED WITH X-VALUE OF POINT TO BE PLOTTED. RETURNS THE
350 REM # OF CHARACTER POSITIONS LEFT OF LEFT MARGIN, OR 1 IF X<=0.
360 REM NEEDS L0=MIN X-VALUE ON PLOT, P7=NICE POSITION WIDTH.

370 DEF FNP(X) = FNI((X - L0) / P7) * SGN( SGN( FNI((X - L0) / P7)) +
                                                          1) + 1

380 REM GAUSSIAN CUMULATIVE APPROXIMATION. PROB FROM -INF TO X.

390 DEF FNG(Z) = SGN( SGN(Z) - 1) + 1 - (2 * SGN( SGN(Z) - 1) + 1) *
                                                  FND( ABS(Z))

400 REM APPROX HALF-GAUSSIAN CUMULATIVE. GOOD TO E-4 FOR 0<=Z<5.5
410 REM REF: DERENZO, MATH. COMP. 31 (1977), 214-225.

420 DEF FND(Z) = EXP( - ((83 * Z + 351) * Z + 562) * Z / (703 + 165 *
                                                  Z)) / 2

430 REM CEILING FUNCTION.

440 DEF FNC(X) = - FNF( - X)
```

```
450 REM   ***DIMENSIONS AND INITIALIZATION***
460 REM TWO DATA ARRAYS, WORK ARRAY, ROW SUBSCRIPTS ARRAY,
470 REM COLUMN SUBSCRIPTS ARRAY, NICE NUMBER ARRAY, AND A PRINT ARRAY.

480 DIM X(200),Y(200),W(211),R(200),C(200),T(30),P(120)

490 REM EPSILON-- 1+E0>1, BUT JUST BARELY. SET E0 ACCORDING TO MACHINE.

500 READ E0
510 DATA   1.0E-06

520 REM PRINTING DETAILS: LEFT MARGIN, RIGHT MARGIN
530 REM TAB(0) SHOULD BE LEFT MARGIN OF PAGE. IF NOT, SET M0>=1.

540 READ M0,M9
550 DATA   0,72

560 REM NICE NUMBERS
570 REM N9 SETS READ SO THAT T(I) POINTS TO THE START OF SET I.

580 READ N9
590 LET K = N9 + 2
600 FOR I = 1 TO N9
610   LET T(I) = K
620   READ J1
630   FOR J = 1 TO J1
640     READ T(K)
650     LET K = K + 1
660   NEXT J
670 NEXT I
680 LET T(N9 + 1) = K
690 DATA   3
700 DATA   3,1,5,10
710 DATA   4,1,2,5,10
720 DATA   9,1,1.5,2,2.5,3,4,5,7,10
730 LET N5 = 2

740 REM VERSION:USUALLY V1=1 IS BRIEF, V1=2 IS VERBOSE
750 REM V1<0 ALLOWS REQUEST FOR USER INPUT (THEREAFTER ABS(V1) USED)

760 LET V1 = 2

770 REM ABOVE INITIALIZATION LINES CAN BE DELETED FOR SPACE
780 REM  GO FROM HERE TO COMMAND-LEVEL.

790 GO TO 4000
```

```
1000 REM SHELL SORT

1010 LET I1 = N - 1
1020 LET I1 = INT((I1 - 2) / 3) + 1
1030 FOR I2 = 1 TO N - I1
1040    LET I0 = I2 + I1
1050    LET W1 = W(I0)
1060    IF W(I2) <= W1 THEN 1140
1070    LET J0 = I2
1080    LET W(I0) = W(J0)
1090    LET I0 = J0
1100    IF J0 < = I1 THEN 1130
1110    LET J0 = J0 - I1
1120    IF W(J0) > W1 THEN 1080
1130    LET W(I0) = W1
1140 NEXT I2
1150 IF I1 > 1 THEN 1020
1160 RETURN

1200 REM    SORT ON X() CARRYING Y()

1210 LET I1 = N - 1
1220 LET I1 = INT((I1 - 2) / 3) + 1
1230 FOR I2 = 1 TO N - I1
1240    LET I0 = I2 + I1
1250    LET X1 = X(I0)
1260    LET Y1 = Y(I0)
1270    IF X(I2) <= X1 THEN 1370
1280    LET J0 = I2
1290    LET X(I0) = X(J0)
1300    LET Y(I0) = Y(J0)
1310    LET I0 = J0
1320    IF J0 < = I1 THEN 1350
1330    LET J0 = J0 - I1
1340    IF X(J0) > X1 THEN 1290
1350    LET X(I0) = X1
1360    LET Y(I0) = Y1
1370 NEXT I2
1380 IF I1 > 1 THEN 1220
1390 RETURN

1400 REM SWAP X() AND Y()

1410 FOR I0 = 1 TO N
1420    LET X1 = X(I0)
1430    LET X(I0) = Y(I0)
1440    LET Y(I0) = X1
1450 NEXT I0
1460 RETURN
```

```
1900 REM SUBROUTINE TO FIND NICE POSITION WIDTH
1910 REM H1,L0=DATA BOUNDS,N5 SELECTS NUMBER SET.P9=DESIRED
1920 REM NUMBER OF POSITIONS,A8=1 IF "-0" OCCURS,ELSE 0
1930 REM ON EXIT: N4=MANTISSA, N3=EXPONENT, U=UNIT=10^N3
1940 REM P8=NUMBER REQUIRED POSITIONS,P7=NICE POSITION WIDTH

1950 IF N5 <= N9 THEN 1980
1960 PRINT TAB(M0);"ILLEGAL N5 IN NPW"
1970 STOP
1980 LET N1 = (H1 - L0) / P9
1990 IF N1 > 0 GO TO 2020
2000 PRINT TAB(M0);"HI <= LO IN NPW"
2010 STOP
2020 LET N3 = FNF( FNL(N1))
2030 LET U = 10 ^ N3
2040 LET N4 = N1 / U
2050 FOR I0 = T(N5) TO T(N5 + 1) - 1
2060    IF N4 <= T(I0) THEN 2090
2070 NEXT I0
2080 LET I0 = T(N5 + 1) - 1
2090 LET N4 = T(I0)
2100 LET P7 = N4 * U

2110 REM COMPUTE NUMBER OF CHARACTER POSITIONS REQUIRED

2120 LET P8 = FNI(H1 / P7) - FNI(L0 / P7) + 1

2130 REM IF -0 POSSIBLE AND (H1 AND L0 HAVE OPPOSITE SIGNS OR H1=0)
2140 REM WE'LL NEED THE -0 LINE

2150 IF A8 = 0 THEN 2210
2160 IF H1 = 0 THEN 2180
2170 IF H1 * (L0 / U) >= 0 THEN 2210
2180 IF P9 = 1 THEN 2220
2190 LET P8 = P8 + 1

2200 REM NOW P8=POSITIONS REQUIRED WITH THIS WIDTH
2210 REM CHECK RANGE COVERED AND ADJUST IF WIDTH IS TOO SMALL

2220 IF P8 <= P9 THEN 2290
2230 LET I0 = I0 + 1
2240 IF I0 <= T(N5 + 1) - 1 THEN 2090
2250 LET I0 = 1
2260 LET U = U * 10
2270 LET N3 = N3 + 1
2280 GO TO 2090
2290 RETURN
```

```
2500 REM SUBROUTINE YINFO TO FIND SUMMARIES FOR N ORDERED VALUES IN W()
2510 REM L1,L2,L3=MEDIAN,LO HINGE, HI HINGE, S1=STEP=1.5*HSPREAD.
2520 REM A3,A4 (A1,A2)=LO AND HI ADJACENT VALUES (THEIR SUBSCRIPTS IN
                                                                W())

2530 IF N >= 3 THEN 2560
2540 PRINT TAB(M0);"N TOO SMALL IN YINFO"
2550 STOP
2560 LET K0 = (N + 1) / 2
2570 LET L1 = FNM(K0)
2580 LET K0 = INT(K0 + 1) / 2
2590 LET K1 = INT(K0)
2600 LET L2 = FNM(K0)
2610 LET L3 = W(N - K1 + 1)
2620 IF K1 = K0 THEN 2640
2630 LET L3 = (L3 + W(N - K1)) / 2
2640 LET S1 = 1.5 * (L3 - L2)
2650 LET F1 = L2 - S1
2660 LET F2 = L3 + S1
2670 FOR A1 = 1 TO K1
2680    IF F1 <= W(A1) THEN 2720
2690 NEXT A1
2700 PRINT TAB(M0);"W()NOT SORTED IN YINFO"
2710 STOP
2720 FOR A2 = N TO N - K1 + 1 STEP - 1
2730    IF F2 >= W(A2) THEN 2760
2740 NEXT A2
2750 GO TO 2700
2760 LET A3 = W(A1)
2770 LET A4 = W(A2)
2780 RETURN
3000 REM SORT X() INTO W() FROM J1 TO J2. USES J1, J2, I1, I
3010 REM ENTRY POINT 1: SORT FROM 1 TO N

3020 LET J1 = 1
3030 LET J2 = N

3040 REM ENTRY POINT 2: SORT FROM J1 TO J2

3050 LET N = J2 - J1 + 1
3060 IF N > 0 THEN 3090
3070 PRINT TAB(M0);"ILLEGAL LIMITS IN COPYSORT"
3080 STOP
3090 LET I1 = 0
3100 FOR I = J1 TO J2
3110    LET I1 = I1 + 1
3120    LET W(I1) = X(I)
3130 NEXT I
3140 GOSUB 1000
3150 RETURN
```

```
3300 REM   SORT Y() INTO W() FROM J1 TO J2.
3310 REM ENTRY POINT 1: SORT FROM 1 TO N

3320 LET J1 = 1
3330 LET J2 = N

3340 REM ENTRY POINT 2: SORT FROM J1 TO J2

3350 LET N = J2 - J1 + 1
3360 IF N > 0 THEN 3390
3370 PRINT TAB(M0);"ILLEGAL LIMITS IN COPYSORT"
3380 STOP
3390 GOSUB 3710
3400 GOSUB 1000
3410 RETURN

3600 REM COPY Y() FROM J1 TO J2 INTO W() STARTING AT I1
3610 REM USES J1,J2,I1,I0.   LEAVES N=J2-J1+1
3640 REM
3650 REM ENTRY HERE COPIES FROM 1 TO N ON BOTH

3660 LET I1 = 1

3670 REM ENTRY HERE COPIES FROM 1 TO N IN Y() STARTS AT I1 IN W()

3680 LET J1 = 1
3690 LET J2 = N

3700 REM ENTRY HERE NEEDS J1,J2,I1 SET

3710 FOR I0 = J1 TO J2
3720    LET W(I1) = Y(I0)
3730    LET I1 = I1 + 1
3740 NEXT I0
3750 RETURN
3800 REM SWAP Y() AND W(), LENGTH N

3810 FOR I0 = 1 TO N
3820    LET X1 = W(I0)
3830    LET W(I0) = Y(I0)
3840    LET Y(I0) = X1
3850 NEXT I0
3860 RETURN
4000 REM SIMPLE DRIVER FOR SMALL INTERPRETER
4010 INPUT Q$
4015 IF Q$ = "AGAIN" THEN 4050
4020 IF Q$ <> "STOP" THEN 4040
4030 STOP

4040 REM <OVERLAY Q$ AT 5000 HOWEVER THE OPERATING SYSTEM ALLOWS>

4050 GOSUB 5000
4060 PRINT
4070 GO TO 4010
```

B. 2 FORTRAN

```
      BLOCK DATA
C
C     CHARS   CONTAINS THE SYMBOLS OF THE STANDARD FORTRAN CHARACTER SET,
C     AND  CHA - CHPT ARE THE CORRESPONDING INDICES INTO CHARS.
C     PUTCHR   IS THE PRIMARY USER OF THIS TRANSLATION VECTOR.
C
      COMMON /CHARIO/ CHARS, CMAX,
     1 CHA, CHB, CHC, CHD, CHE, CHF, CHG, CHH, CHI, CHJ, CHK,
     2 CHL, CHM, CHN, CHO, CHP, CHQ, CHR, CHS, CHT, CHU, CHV,
     3 CHW, CHX, CHY, CHZ, CHO, CH1, CH2, CH3, CH4, CH5, CH6,
     4 CH7, CH8, CH9, CHBL, CHEQ, CHPLUS, CHMIN, CHSTAR, CHSLSH,
     5 CHLPAR, CHRPAR, CHCOMA, CHPT
C
C
      INTEGER CHARS(46), CMAX
      INTEGER CHA, CHB, CHC, CHD, CHE, CHF, CHG, CHH, CHI
      INTEGER CHJ, CHK, CHL, CHM, CHN, CHO, CHP, CHQ, CHR
      INTEGER CHS, CHT, CHU, CHV, CHW, CHX, CHY, CHZ
      INTEGER CHO, CH1, CH2, CH3, CH4, CH5, CH6, CH7, CH8, CH9
      INTEGER CHBL, CHEQ, CHPLUS, CHMIN, CHSTAR, CHSLSH
      INTEGER CHLPAR, CHRPAR, CHCOMA, CHPT
C
      DATA CHARS( 1),CHARS( 2),CHARS( 3),CHARS( 4) /1HA,1HB,1HC,1HD/
      DATA CHARS( 5),CHARS( 6),CHARS( 7),CHARS( 8) /1HE,1HF,1HG,1HH/
      DATA CHARS( 9),CHARS(10),CHARS(11),CHARS(12) /1HI,1HJ,1HK,1HL/
      DATA CHARS(13),CHARS(14),CHARS(15),CHARS(16) /1HM,1HN,1HO,1HP/
      DATA CHARS(17),CHARS(18),CHARS(19),CHARS(20) /1HQ,1HR,1HS,1HT/
      DATA CHARS(21),CHARS(22),CHARS(23),CHARS(24) /1HU,1HV,1HW,1HX/
      DATA CHARS(25),CHARS(26),CHARS(27),CHARS(28) /1HY,1HZ,1HO,1H1/
      DATA CHARS(29),CHARS(30),CHARS(31),CHARS(32) /1H2,1H3,1H4,1H5/
      DATA CHARS(33),CHARS(34),CHARS(35),CHARS(36) /1H6,1H7,1H8,1H9/
      DATA CHARS(37),CHARS(38),CHARS(39),CHARS(40) /1H ,1H=,1H+,1H-/
      DATA CHARS(41),CHARS(42),CHARS(43),CHARS(44) /1H*,1H/,1H(,1H)/
      DATA CHARS(45),CHARS(46)                     /1H,,1H./
      DATA CMAX /46/
      DATA CHA,CHB,CHC,CHD,CHE,CHF          / 1, 2, 3, 4, 5, 6/
      DATA CHG,CHH,CHI,CHJ,CHK,CHL          / 7, 8, 9,10,11,12/
      DATA CHM,CHN,CHO,CHP,CHQ,CHR          /13,14,15,16,17,18/
      DATA CHS,CHT,CHU,CHV,CHW,CHX          /19,20,21,22,23,24/
      DATA CHY,CHZ,CHO,CH1,CH2,CH3          /25,26,27,28,29,30/
      DATA CH4,CH5,CH6,CH7,CH8,CH9          /31,32,33,34,35,36/
      DATA CHBL,CHEQ,CHPLUS,CHMIN           /37,38,39,40/
      DATA CHSTAR,CHSLSH,CHLPAR,CHRPAR      /41,42,43,44/
      DATA CHCOMA,CHPT                      /45,46/
C
C
      END
```

```
      SUBROUTINE CINIT(IOUNIT, IPMIN, IPMAX, IEPSI, IMAXIN, ERR)
C
      INTEGER IOUNIT, IPMIN, IPMAX, IMAXIN, EPR
      REAL IEPSI
C
C   INITIALIZATION, TO BE CALLED AT START OF ANY  MAIN PROGRAM
C   WHICH CALLS ONE OF THE EDA SUBROUTINES (EITHER DIRECTLY OR
C   INDIRECTLY).
C
C   IOUNIT  IS THE NUMBER OF THE UNIT TO WHICH OUTPUT IS DIRECTED.
C   IPMIN   IS THE LEFT MARGIN.
C   IPMAX   IS THE RIGHT MARGIN.
C   IEPSI   IS THE MACHINE-RELATED EPSILON.
C   IMAXIN  IS THE MAXIMUM PERMITTED INTEGER VALUE
C
C   ERR IS THE (USUAL) ERROR FLAG, TO INDICATE WHETHER
C   THE ROUTINE EXECUTED SUCCESSFULLY.
C
      COMMON /CHRBUF/ P, PMAX, PMIN, OUTPTR, MAXPTR, OUNIT
      COMMON /NUMBRS/ EPSI, MAXINT
C
      INTEGER P(130), PMAX, PMIN, OUTPTR, MAXPTR, OUNIT
      REAL EPSI, MAXINT
C
C   LOCAL VARIABLES
C
      INTEGER BLANK, I
      DATA BLANK /1H /
C
C
      ERR = 6
      IF(IPMIN .LT. 1) GO TO 999
      IF(IPMAX .GT. 130) GO TO 999
      IF(IPMAX .LE. IPMIN) GO TO 999
      ERR = 7
      IF((1.0 + IEPSI) .LE. 1.0) GO TO 999
      ERR = 0
      OUNIT = IOUNIT
      PMIN = IPMIN
      OUTPTR = IPMIN
      MAXPTR = IPMIN
      PMAX = IPMAX
      EPSI = IEPSI
      MAXINT = FLOAT(IMAXIN)
C
      DO 50 I = 1, 130
        P(I) = BLANK
   50 CONTINUE
C
  999 RETURN
      END
```

```
      SUBROUTINE PUTCHR(POSN, CHAR, ERR)
C
      INTEGER POSN, CHAR, ERR
C
C PLACE THE CHARACTER  CHAR  AT POSITION  POSN   IN
C THE OUTPUT LINE  P .  IF  POSN = 0 , PLACE  CHAR  IN THE
C NEXT AVAILABLE POSITION IN  P .  MAXPTR  IS TO BE INITIAL-
C IZED TO  PMIN , AND  PRINT  MUST RESET IT.
C
      COMMON /CHARIO/ CHARS, CMAX,
     1 CHA, CHB, CHC, CHD, CHE, CHF, CHG, CHH, CHI, CHJ, CHK,
     2 CHL, CHM, CHN, CHO, CHP, CHQ, CHR, CHS, CHT, CHU, CHV,
     3 CHW, CHX, CHY, CHZ, CH0, CH1, CH2, CH3, CH4, CH5, CH6,
     4 CH7, CH8, CH9, CHBL, CHEQ, CHPLUS, CHMIN, CHSTAR, CHSLSH,
     5 CHLPAR, CHRPAR, CHCOMA, CHPT
C
      COMMON /CHRBUF/ P, PMAX, PMIN, OUTPTR, MAXPTR, OUNIT
C
      INTEGER CHARS(46), CMAX
      INTEGER CHA, CHB, CHC, CHD, CHE, CHF, CHG, CHH, CHI
      INTEGER CHJ, CHK, CHL, CHM, CHN, CHO, CHP, CHQ, CHR
      INTEGER CHS, CHT, CHU, CHV, CHW, CHX, CHY, CHZ
      INTEGER CH0, CH1, CH2, CH3, CH4, CH5, CH6, CH7, CH8, CH9
      INTEGER CHBL, CHEQ, CHPLUS, CHMIN, CHSTAR, CHSLSH
      INTEGER CHLPAR, CHRPAR, CHCOMA, CHPT
      INTEGER P(130), PMAX, PMIN, OUTPTR, MAXPTR, OUNIT
C
      IF(CHAR .GT. 0 .AND. CHAR .LE. CMAX) GO TO 10
      ERR = 4
      RETURN
   10 IF(POSN .NE. 0) OUTPTR = MAXO(PMIN, POSN)
      OUTPTR = MINO(OUTPTR, PMAX)
      P(OUTPTR) = CHARS(CHAR)
      MAXPTR = MAXO(MAXPTR, OUTPTR)
      OUTPTR = OUTPTR + 1
      RETURN
      END
      INTEGER FUNCTION WDTHOF(I)
      INTEGER I
C FIND THE NUMBER OF CHARACTERS NEEDED TO PRINT I
      INTEGER IA, IQ, ND
C
      IA = IABS(I)
      ND = 1
      IF(I .LT. 0) ND = 2
   10 IQ = IA/10
      IF(IQ .EQ. 0) GO TO 20
        IA = IQ
        ND = ND + 1
        GO TO 10
   20 WDTHOF = ND
      RETURN
      END
```

```
       SUBROUTINE PUTNUM(POSN, N, W, ERR)
C
       INTEGER POSN, N, W, ERR
C
C  PLACE THE CHARACTER REPRESENTATION OF THE INTEGER  N
C  RIGHT-JUSTIFIED IN A FIELD  W  SPACES WIDE STARTING
C  AT POSITION  POSN  IN THE OUTPUT LINE  P .
C
C  THE VARIABLES  IP, INUM, AND  IW  ARE INTERNAL VERSIONS
C  OF  POSN, N, AND  W .  WE PROCEED BY EXTRACTING THE
C  DIGITS OF  N, STARTING WITH THE LOW-ORDER DIGIT,
C  AND STACKING THEM IN  DSTK. ( ND  COUNTS THE DIGITS.)
C  ONCE WE HAVE COLLECTED ALL THE DIGITS (AND KNOW THAT
C   W  SPACES ARE SUFFICIENT), WE SKIP OVER ANY UNNEEDED
C  SPACES, PUT OUT A MINUS SIGN IF NEEDED, AND THEN PUT OUT
C  THE DIGITS, STARTING WITH THE HIGH-ORDER ONE.
C
C  THIS ROUTINE CALLS PUTCHR   AND DEPENDS ON HAVING DIGITS
C  0 THROUGH 9 IN CONSECUTIVE ELEMENTS OF  CHARS  IN THE
C  COMMON BLOCK  CHARIO, STARTING AT  CH0 = 27.  IT ALSO
C  ASSUMES THAT THE MINUS SIGN IS AT  CHMIN = 40 IN  CHARS.
C
       INTEGER CHD, CH0, CHMIN, DSTK(20), INUM, IP, IQ, IW, ND
C
       COMMON/CHRBUF/ P, PMAX, PMIN, OUTPTR, MAXPTR, OUNIT
       INTEGER P(130), PMAX, PMIN, OUTPTR, MAXPTR, OUNIT
C
       DATA CH0, CHMIN/27, 40/
C
C
       IW = W
       IF(N .LT. 0) IW = IW - 1
       INUM = IABS(N)
C
C  EXTRACT AND STACK THE DIGITS OF  INUM, CHECKING
C  TO SEE THAT  N  FITS IN  W  SPACES.
C
       ND = 1
   10  IQ = INUM/10
       DSTK(ND) = INUM - IQ * 10
       IF(ND .LE. 20 .AND. ND .LE. IW) GO TO 20
         ERR = 2
         GO TO 999
   20  IF(IQ .EQ. 0) GO TO 30
         INUM = IQ
         ND = ND + 1
         GO TO 10
C
C  UNSTACK THE DIGITS FROM  DSTK  AND PUT THEM OUT.
C  NOTE THAT WHEN  N  IS NEGATIVE, A MINUS SIGN MUST BE
C  INSERTED IN THE SPACE BEFORE THE FIRST DIGIT.  DECREASING
C   IW  BY 1  IN THE INITIALIZATION HAS PROVIDED A SPACE
C  FOR THE MINUS SIGN.
```

```
C
   30 IP = POSN
      IF(IP .EQ. 0) IP = OUTPTR
      IP = IP + IW - ND
      IF(N .GE. 0) GO TO 40
        CALL PUTCHR(IP, CHMIN, ERR)
        IP = IP + 1
   40 CHD = CH0+ DSTK(ND)
      CALL PUTCHR(IP, CHD, ERR)
      IF(ND .EQ. 1) GO TO 50
        ND = ND - 1
        IP = IP + 1
        GO TO 40
   50 CONTINUE
C
  999 RETURN
      END
```


```
      SUBROUTINE PRINT
C
C PRINT THE OUTPUT LINE  P  ON UNIT  OUNIT  (MAXPTR
C INDICATES THE RIGHTMOST POSITION WHICH HAS BEEN USED
C IN THIS LINE).  THEN RESET  P  TO SPACES, AND  MAXPTR  AND
C OUTPTR TO PMIN.
C
      COMMON /CHRBUF/ P, PMAX, PMIN, OUTPTR, MAXPTR, OUNIT
C
      INTEGER P(130), PMAX, PMIN, OUTPTR, MAXPTR, OUNIT
C
C LOCAL VARIABLES
C
      INTEGER  BLANK, I
C
      DATA BLANK /1H /
C
      WRITE(OUNIT, 10) (P(I), I=1, MAXPTR)
   10 FORMAT(1X, 130A1)
C
      DO 20 I = 1, MAXPTR
        P(I) = BLANK
   20 CONTINUE
C
      OUTPTR = PMIN
      MAXPTR = PMIN
C
      RETURN
      END
```

```
      SUBROUTINE SORT( Y, N, ERR)
C
      INTEGER N, ERR
      REAL Y(N)
C
C SHELL SORT  N  VALUES IN  Y()  FROM SMALLEST TO LARGEST.
C
C NOTE THAT LOCAL SYSTEM SORT UTILITIES ARE LIKELY TO BE
C MORE EFFICIENT, AND SHOULD BE SUBSTITUTED WHENEVER POSSIBLE.
C
C LOCAL VARIABLES
C
      INTEGER I, J, J1, GAP, NMG
      REAL TEMP
C
      IF(N .GE. 1) GO TO 10
        ERR = 1
        GO TO 999
   10 IF(N .EQ. 1) GO TO 999
C
C ONE ELEMENT IS ALWAYS SORTED
C
      GAP = N
   20 GAP = GAP/2
      NMG = N - GAP
      DO 40 J1 = 1, NMG
        I = J1 + GAP
C
C DO  J = J1, 1, -GAP
C
        J = J1
   30   IF (Y(J) .LE. Y(I)) GO TO 40
C
C SWAP OUT-OF-ORDER PAIR
C
        TEMP = Y(I)
        Y(I) = Y(J)
        Y(J) = TEMP
C
C KEEP OLD POINTER FOR NEXT TIME THROUGH
C
        I = J
        J = J - GAP
        IF (J .GE. 1) GO TO 30
   40 CONTINUE
      IF (GAP .GT. 1) GO TO 20
  999 RETURN
      END
```

```
       SUBROUTINE PSORT( ON, WITH, N, ERR)
C
       INTEGER N, ERR
       REAL ON(N), WITH(N)
C
C  PAIR SHELL SORT  N  VALUES IN  ON()  FROM SMALLEST TO LARGEST
C  CARRYING ALONG THE VALUES IN  WITH().
C
C  NOTE THAT LOCAL SYSTEM SORT UTILITIES ARE LIKELY TO BE
C  MORE EFFICIENT, AND SHOULD BE SUBSTITUTED WHENEVER POSSIBLE.
C
C  LOCAL VARIABLES
C
       INTEGER I, J, J1, GAP, NMG
       REAL TON,TWITH
C
       IF(N .GE. 1) GO TO 10
         ERR = 1
         GO TO 999
   10 IF(N .EQ. 1) GO TO 999
C
C  ONE ELEMENT IS ALWAYS SORTED
C
       GAP = N
   20 GAP = GAP/2
       NMG = N - GAP
       DO 40 J1 = 1, NMG
         I = J1 + GAP
C
C  DO   J = J1, 1, -GAP
C
       J = J1
   30    IF (ON(J) .LE. ON(I)) GO TO 40
C
C  SWAP OUT-OF-ORDER PAIR
C
       TON = ON(I)
       ON(I) = ON(J)
       ON(J) = TON
       TWITH = WITH(I)
       WITH(I) = WITH(J)
       WITH(J) = TWITH
C
C  KEEP OLD POINTER FOR NEXT TIME THROUGH
C
       I = J
       J = J - GAP
       IF (J .GE. 1) GO TO 30
   40 CONTINUE
       IF (GAP .GT. 1) GO TO 20
  999 RETURN
       END
```

```
         SUBROUTINE YINFO(Y, N, MED, HL, HH, ADJL, ADJH, IADJL, IADJH,
     1   STEP, ERR)
C
C   GET GENERAL INFORMATION ABOUT Y().  USEFUL FOR PLOT SCALING.
C   SORTS Y() AND RETURNS IT SORTED.  ALSO RETURNS
C       MED =  MEDIAN
C       HL  =  LOW HINGE            HH   =HI HINGE
C       ADJL = LOW ADJACENT VALUE  ADJH =HI ADJ VALUE
C       IADJL=  ITS INDEX (LOCATN) IADJH=ITS INDEX
C
         INTEGER N, IADJL, IADJH, ERR
         REAL Y(N), MED, HL, HH, ADJL, ADJH, STEP
C
C   LOCAL VARIABLES
C
         REAL  HFENCE, LFENCE
         INTEGER J, K, TEMP1, TEMP2
C
         CALL SORT(Y, N, ERR)
         IF (ERR .NE. 0) GO TO 999
         K=N
         J = (K/2)+1
C
         TEMP1 = N+1-J
         MED = (Y(J) + Y(TEMP1))/2.0
C
         K = (K+1)/2
         J = (K/2) + 1
         TEMP1 = K+1-J
         HL = (Y(J) + Y(TEMP1))/2.0
         TEMP1 = N-K+J
         TEMP2 = N+1-J
         HH = (Y(TEMP1) + Y(TEMP2))/2.0
C
         STEP = (HH - HL)*1.5
         HFENCE = HH + STEP
         LFENCE = HL - STEP
C
C   FIND ADJACENT VALUES
C
         IADJL = 0
   20    IADJL = IADJL + 1
         IF ( Y(IADJL) .LE. LFENCE) GO TO 20
         ADJL = Y(IADJL)
C
         IADJH = N+1
   30    IADJH = IADJH - 1
         IF ( Y(IADJH) .GE. HFENCE) GO TO 30
         ADJH = Y(IADJH)
  999    RETURN
         END
```

```
          SUBROUTINE NPOSW(HI, LO, NICNOS, NN, MAXP, MZERO, PTOTL, FRACT,
        1  UNIT, NPW, ERR)
C
C    FIND A NICE (I.E., SIMPLE) DATA-UNITS VALUE TO ASSIGN TO ONE PLOT
C    POSITION IN ONE DIMENSION OF A PLOT.  A PLOT POSITION IS TYPICALLY
C    ONE CHARACTER POSITION HORIZONTALLY, OR ONE LINE VERTICALLY.
C
C    ON ENTRY:
C    HI, LO  ARE THE HIGH AND LOW EDGES OF THE DATA RANGE TO BE PLOTTED.
C    NICNOS  IS A VECTOR OF LENGTH  NN  CONTAINING NICE MANTISSAS FOR
C            THE PLOT UNIT.
C    MAXP    IS THE MAXIMUM NUMBER OF PLOT POSITIONS ALLOWED IN THIS
C            DIMENSION OF THE PLOT.
C    MZERO   IS .TRUE. IF A POSITION LABELED  -0  US ALLOWED IN THIS
C            DIMENSION,  .FALSE. OTHERWISE.
C
C    ON EXIT:
C    PTOTL   HOLDS THE TOTAL NUMBER OF PLOT POSITIONS TO BE USED IN
C            THIS DIMENSION.  (MUST BE .LE. MAXP.)
C    FRACT   IS THE MANTISSA OF THE NICE POSITION WIDTH.  IT IS
C            SELECTED FROM THE NUMBERS IN NICNOS.
C    UNIT    IS AN INTEGER POWER OF 10 SUCH THAT NPW = FRACT * UNIT.
C    NPW     IS THE NICE POSITION WIDTH.  ONE PLOT POSITION WIDTH
C            WILL REPRESENT  A DATA-SPACE DISTANCE OF  NPW.
C
C
          INTEGER NN, MAXP, PTOTL, ERR
          REAL HI, LO, NICNOS(NN), FRACT, UNIT, NPW
          LOGICAL MZERO
C
C    FUNCTIONS
          INTEGER FLOOR, INTFN
C
C    LOCAL VARIABLES
C
          INTEGER I
          REAL APRXW
C
          IF (MAXP .GT. 0) GO TO 5
          ERR = 8
          GO TO 999
        5 APRXW = (HI - LO)/FLOAT(MAXP)
          IF(APRXW .GT. 0.0) GO TO 10
C
C    HI .LE. LO IS AN ERROR
C
          ERR = 9
          GO TO 999
       10 UNIT = 10.0**FLOOR(ALOG10(APRXW))
          FRACT = APRXW/UNIT
          DO 20 I = 1, NN
            IF(FRACT .LE. NICNOS(I)) GO TO 30
       20 CONTINUE
```

```
   30 FRACT = NICNOS(I)
      NPW = FRACT * UNIT
      PTOTL = INTFN(HI/NPW, ERR) - INTFN(LO/NPW, ERR) + 1
      IF(ERR .NE. 0) GO TO 999
C
C  IF MINUS ZERO POSITION POSSIBLE AND SGN(HI) .NE. SGN(LO), ALLOW IT.
C
      IF(MZERO .AND. (HI*LO .LT. 0.0 .OR. HI .EQ. 0.0)) PTOTL=PTOTL+1
C
C  PTOTL POSITIONS REQUIRED WITH THIS WIDTH -- FEW ENOUGH?
C
      IF(PTOTL .LE. MAXP) GO TO 999
C
C  TOO MANY POSITIONS NEEDED, SO BUMP NPW UP ONE NICE NUMBER
C
      I = I+1
      IF(I .LE. NN) GO TO 30
      I = 1
      UNIT = UNIT * 10.0
      GO TO 30
  999 RETURN
      END




      INTEGER FUNCTION INTFN(X, ERR)
C
C  FIND THE INTEGER EQUAL TO OR NEXT CLOSER TO ZERO THAN X.
C
C  CHECKS TO SEE THAT  X  IS NOT TOO LARGE TO FIT IN AN
C  INTEGER VARIABLE.
C
      REAL X
      INTEGER ERR
C
      COMMON /NUMBRS/ EPSI, MAXINT
      REAL EPSI, MAXINT
C
      IF( ABS(X) .LE. MAXINT) GO TO 10
C
C  X  IS TOO LARGE IN MAGNITUDE TO FIT IN AN INTEGER,
C  RETURN THE LARGEST LEGAL INTEGER AND SET THE ERROR FLAG.
C
      ERR = 3
      INTFN = IFIX( SIGN(MAXINT, X) )
      GO TO 999
C
   10 INTFN = INT((1.0 + EPSI) * X)
  999 RETURN
      END
```

```
      INTEGER FUNCTION FLOOR (Y)
      REAL Y
C  FIND FLOOR(Y), THE LARGEST INTEGER NOT EXCEEDING Y
C
      FLOOR = INT(Y)
      IF(Y .LT. 0.0 .AND. Y .NE. FLOAT(FLOOR)) FLOOR = FLOOR - 1
      RETURN
      END



      REAL FUNCTION MEDIAN(Y, N)
C  FIND THE MEDIAN OF THE SORTED VALUES Y(1), ..., Y(N).
      INTEGER N
      REAL Y(N)
C  LOCAL VARIABLES
      INTEGER MPTR, MPT2
C
      MPTR = (N/2) + 1
      MPT2 = N-MPTR+1
      MEDIAN = (Y(MPTR) + Y(MPT2))/2.0
      RETURN
      END



      REAL FUNCTION GAU(Z)
      REAL Z
C  THIS FUNCTION CALCULATES THE VALUE OF THE STANDARD
C  GAUSSIAN CUMULATIVE DISTRIBUTION FUNCTION AT  Z.
C  THE ALGORITHM USES APPROXIMATIONS GIVEN BY STEPHEN E. DERENZO
C  IN MATHEMATICS OF COMPUTATION, V. 31 (1977), PP. 214-225
C
C  LOCAL VARIABLES
      REAL P, PI, X
C
      X = ABS(Z)
      IF(X .GT. 5.5) GO TO 10
      P = EXP(-((83.0 * X + 351.0) * X + 562.0) * X /
    1    (703.0 + 165.0 * X))
      GO TO 20
C
   10 PI = 4.0 * ATAN(1.0)
      P = SQRT(2.0/PI) * EXP(-(X * X/2.0 +
    1    0.94/(X * X))) / X
C
C  THE APPROXIMATIONS YIELD VALUES OF THE HALF-NORMAL TAIL AREA.
C  TRANSLATE THAT INTO THE VALUE OF THE GAUSSIAN C.D.F. AND
C  ALLOW FOR THE SIGN OF Z.
C
   20 GAU = P/2.0
      IF(Z .GT. 0.0) GAU = 1.0 - GAU
C
      RETURN
      END
```

Appendix C

Programming Conventions

The programs in this book form two sets of routines, one in BASIC and one in FORTRAN. This appendix discusses the structure and language conventions adopted for these programs. The first part of the appendix covers the BASIC programs. The second part deals with the FORTRAN programs.

C.1   BASIC

Environment

The BASIC programs in this book are written to run conveniently on computers using an interactive BASIC interpreter. In particular, most mini- and microcomputers should accept these programs with only minor modifications. Users of systems where BASIC is compiled rather than interpreted may have to write a driver program to facilitate interprogram communication. This part of the appendix discusses the structure and conventions of the BASIC programs and provides advice and guidelines for modifying the programs to suit different computing environments.

In many implementations of BASIC, all variables are global and can be modified and manipulated interactively by the user. The list of variable-naming conventions in this section will enable users to take full advantage of this feature. The complete set of programs is between 40K and 50K characters long. However, the programs are organized into a segment of utility subroutines and nine EDA subroutines. With some sort of mass storage under program control (a tape or floppy disk is fine) and an OVERLAY instruction (or DELETE and APPEND on some systems), each EDA routine can be brought into core, used, and then replaced by another in turn. Without this flexibility, individual programs can still be run in little memory, but it will be more difficult to move among them while analyzing data. A sample elementary driver is included for illustration (starting at line number 4000 in Appendix B). Systems with a CHAIN instruction can use it for interprogram linkage, but programmers will need to pay attention to the communication of variable values among routines.

The longest programs require about 12K bytes (characters) of core memory plus room for data (16K is practical, and 24K is comfortable). Hints on trading space for processing time appear later in this appendix.

Program Structure

The programs have the following structure:

Line Nos.	Contents	Comments
10–90	Remarks	Can be used for special control functions such as user-defined keys on some computers.
100–490	Function definitions	Some systems do not permit OVERLAY of function definitions, so they come here.
500–800	Main initialization	This could be a subroutine, but some systems do not permit OVERLAY of data statements.

Line Nos.	Contents	Comments
1000–4000	Utility subroutines	Such operations as sorting and plot scaling
4000–4900	Driver program	A sample elementary driver is included for illustration.
5000–	EDA subroutines	All the EDA programs are written as subroutines which start at line 5000. An OVERLAY 5000 instruction (or its equivalent) is one possible way to bring them into core.

Conventions

We have observed the following variable-naming conventions:

X( ), Y( ) Vectors of length N, hold data. Y( ) is the "dependent" variable and is most often analyzed.

W( ) Workspace vector of length N + 11 (the extra eleven locations are for the smoothing programs).

R( ), C( ) Vectors to hold row and column subscripts, respectively. Some routines use R( ) and C( ) for extra storage or return residuals in R( ).

T( ) Internal vector, holds "nice numbers" for plot scaling.

P( ) Print vector, holds one output line of characters.

E0 Machine epsilon (see *Epsilonics* below).

M0, M9 Left and right margins—TAB(M0) positions the cursor at left margin.

V1 Version number (to select among versions of an analysis or display). Generally V1 = 1 calls for the shortest printout, starkest display, or simplest analysis; larger values of V1 call for more complicated versions. A negative value of V1 signals that the user will supply parameters interactively.

Whenever possible, work is done in W( ), and X( ) and Y( ) are preserved or only reordered. The design philosophy of the BASIC programs has favored minimizing the space required for the storage of data. At times this requires that X( ) and Y( ) be destroyed or used to return a result. On systems with no constraints on storage, extra arrays to preserve X( ) and Y( ) would be valuable and could easily be introduced.

Space versus Speed

The most expensive operation commonly performed by these programs is sorting. Users of microcomputers may find the sorting process noticeably slow. A machine–language sorting program will significantly extend the size of data batches that can be conveniently analyzed. Programmers who wish to optimize this code for a specific machine should first provide a fast sorting program. No other optimization will have nearly as great an effect.

To save space, programs may delete lines 480–790 after they have been executed. Or, if permitted, initialization can be made a subroutine at line 5000 to be called first. Also, most of the EDA subroutines (and all the long subroutines) can be split into two or more segments to be executed in sequence. Thus, for example, plot options could be checked in one program segment; then a second segment could determine plot scaling; and finally, a third segment could produce the plot.

Epsilonics

The decimal numbers with which humans customarily work cannot generally be represented exactly in the binary (or, sometimes, hexadecimal) forms used by most computers. For example, when written as a binary fraction, the number 1/10 is a repeating fraction (.000110011 . . . in binary digits). Because computers store real numbers in fixed-length words, their internal representation will usually be only a very close approximation to the true number. For example, LOG(1000) may be slightly different from 3.0. The *representation errors* that occur in converting decimal numbers to binary and the *rounding errors* that arise in subsequent arithmetic have a negligible effect on most EDA calculations, but there are important exceptions. One of these is the floor operation (the INT function in BASIC; see *Rounding Functions,* below), used especially in scaling plots and placing characters precisely for displays. For example, INT(2.9999) yields 2.0. Thus, because LOG(1000) may not be represented as exactly 3, INT(LOG(1000)) might come out 2 rather than 3. If we do not allow for these errors, small as they may be, many programs will run into serious (and obscure) trouble. To correct this problem, we introduce a machine-dependent constant, epsilon (E0 in the BASIC programs), which is the smallest number such that (in the computer's arithmetic) $1.0 + \epsilon > 1.0$. We use a slightly larger number for E0. (1.0 E−6 works well on most machines which use 4 bytes to hold a number.) If E0 is too small, many anomalous things can happen, including incorrect stem-and-leaf displays and *x-y* plots.

Some BASIC implementations provide a user-adjusted "fuzz" factor that will accomplish a similar function in computations. This feature may be able to replace the epsilon in the defined functions FNF and FNI.

BASIC Portability

The BASIC programs in this book are written in a dialect of BASIC as close to the ANSI minimal BASIC standard as possible. Since few BASIC implementations are in fact ANSI-standard, we note here some specific features that may require the attention of a programmer when installing these programs. (Our reference for some of these notes is "BASIC REVISITED, An Update to Interdialect Translatability of the BASIC Programming Language" by Gerald L. Isaacs, CONDUIT, University of Iowa, 1976.)

Variable Names. BASIC variable names are single letters or a single letter followed by a single digit. Some implementations of BASIC permit longer variable names, but a program using longer names would not be portable. We have deliberately made some variable names mnemonic. Thus L0 (L-zero) and H1 (H-one) often hold the low and high data values of a batch. String variables obey the same rules and end in $. We have restricted array names to single letters. This is less general than the ANSI standard but required by some BASIC implementations.

String Functions. Three string-related functions not in the ANSI standard are used throughout the programs for displays. These are

LEN(A$)   The number of characters in the string A$.
STR$(N)   The numerals representing the number N. This

function is needed to produce a numeral with no blank spaces before or after it. One possible substitute is a subroutine that constructs the numeral string by selecting characters from a string array or from the string "0123456789" by using a substring operation.

ASC("C")   The ASCII code value of the character "C". This function is used for ease of exposition and can easily be replaced by the literal numeric value. Non-ASCII systems should use the appropriate character codes.

Some of these functions have different names on some systems.

String Variables.   The programs occasionally use string variables and string constants. String constants are enclosed in double quotes ("). Numeric codes can be substituted for many, but not all, string uses.

Loops.   FOR loops are supposed to check the index variable at the top of the loop. Thus FOR I = 10 TO 9 STEP 1 should skip the loop entirely (rather than executing it once). Some versions of BASIC test the index variable at the end of the loop instead. We have, therefore, provided special checks when necessary before loops. Similarly, index variables are not defined reliably at the end of loops. We have inserted an assignment statement after some loops to ensure that the index variable is set correctly.

Margins.   The left margin, M0, is usually set to zero. In some versions of BASIC, TAB(0) is not the same as the first print position, so M0 may need to be set to 1.

Defined Functions.   The programs include several user-defined functions, but one-line defined functions are sufficient, provided that a defined function can use a previously defined function in its definition. The ANSI standard requires a single argument for defined functions and global access to all variables. If multiple-line or multiple-argument defined functions are available, programmers may wish to modify some of the functions for greater efficiency and clarity.

Rounding Functions.   The programs require a function that returns the largest integer not exceeding its argument. This is commonly known as the "floor function," but it is called INT in BASIC. Rounding functions can be a source of great confusion (and subtle bugs). We might round a number in four ways, as shown in the table.

Name	Rounding Direction	Symbol	Result for $x =$ 2.4	-2.4
floor	down	$\lfloor x \rfloor$	2	-3
int(eger part)	in, toward zero	$[x]$	2	-2
ceiling	up	$\lceil x \rceil$	3	-2
"outt"	out, away from zero	$]x[$	3	-3

The "outt" function is rarely discussed (and our name and notation for it are fanciful), but the operation is used in these programs to set display boundaries to the next integer value outside some bounds. Each rounding operation could include some epsilonics (as discussed earlier) to avoid problems introduced by representation and rounding errors. Each of these functions can be defined in one line from some of the others (plus the absolute value function, abs(x), and the signum function, sgn(x), which returns $+1$, 0, or -1 when x is positive, zero, or negative, respectively); for example:

$$\text{floor}(x) = \text{int}(x) + \text{sgn}(\text{sgn}(x - \text{int}(x)) + 1) - 1$$
$$\text{int}(x) = \text{sgn}(x) * \text{floor}(\text{abs}(x))$$
$$\text{ceiling}(x) = - \text{floor}(-x)$$
$$\text{outt}(x) = \text{sgn}(x) * \text{ceiling}(\text{abs}(x))$$

Note again that INT(X) in BASIC is floor(X).

Errors. Because the BASIC programs will usually run interactively, they report errors immediately and stop execution. When the programs are run on an interpreter, the user will have a chance to correct the error and restart from that point.

C.2 FORTRAN

We hardly need to explain our decision to provide programs in FORTRAN— it is the most nearly universal of all scientific programming languages. We cannot, however, pretend that developing these programs was a labor of love. A reader who examines them carefully will find segments that are awkward or tedious because FORTRAN is ill-suited to the programming needs of modern data analysis. For example, the output capabilities of FORTRAN are far too rigid for the graphic and semi-graphic displays that are common in exploratory data analysis. On the whole, however, the advantages of making these programs as widely available as possible outweighed the difficulties of FORTRAN.

If programs are to be widely used, they must be portable. That is, it must be possible to move them from one computing environment to another with an absolute minimum number of changes. Fortunately for us, others have laid substantial groundwork in developing portable (or, strictly speaking, semi-portable) FORTRAN programs. As a result, a number of practices that facilitate portability are well-established, and computer software to support the most valuable of them is available. In this part of the appendix we briefly describe the practices we have followed and the role they have played in the development of our programs.

Consistency of style is also important for any set of programs that are intended to be used (and *read*) together. Thus we also describe the particular conventions we have chosen to follow. These range from simple choices that affect only the appearance of the printed programs to overall decisions that affect the structure and interrelations among all the programs in this book.

Related to interconnections is the question of just how one might customarily use these programs. We briefly discuss and illustrate two approaches to this.

And finally there are the utility routines, which perform a variety of essential services for the data analysis routines presented in Chapters 1 through 9. Listings for the utility routines appear in Appendix B.

Portability

A fully portable program or subroutine can be moved gracefully from one computing machine to another. And even though the computers are of different manufacture and have different systems software, the program compiles without errors, executes without errors, and produces identically the same results on both. This is the ideal situation. Unfortunately, it can rarely be attained in practice; but with reasonable effort a good approximation to it is possible. The two primary obstacles to overcome are differences among dialects of the FORTRAN language and differences in characteristics of the arithmetic hardware. (One must also contend with variations in system conventions, but these are generally less serious.)

The solution to the problem of dialects is conceptually quite simple: One uses only a subset of FORTRAN that is handled in the same way by essentially all known systems. In practice it is all too easy to slip back unknowingly into using some facility or construction which is acceptable in one's own environment but unacceptable in certain others. To avoid this, we have restricted our FORTRAN to a particular subset known as PFORT. This is an attractive solution because this subset of FORTRAN is supported by a piece of software, the PFORT Verifier (Ryder 1974), that takes a FORTRAN program as input and reports on all its departures from this subset of the language. Especially valuable is the Verifier's ability to process a main program and all associated subroutines and to identify potential difficulties of communication among them, including misuse of COMMON.

When a particular construction is acceptable in many (but not all) dialects of FORTRAN, it is tempting to use it—especially when it would make the programs easier to understand—and then to announce, "The programs conform to PFORT, except for. . . ." For example, subscript expressions of the form $N + 1 - I$ are common (as in LVALS, MEDPOL, and RGCOMP), but the strict FORTRAN definition of subscript expressions is too restrictive to permit this form. We have decided to avoid such complications and adhere to PFORT. Thus we can state that *all the FORTRAN programs in this book have been processed by the PFORT Verifier without any warning messages.*

The problem of arithmetic hardware characteristics is somewhat more difficult than the problem of language dialects. Fortunately, EDA techniques generally involve much less numerical computation than one finds in most mathematical software. In fact, our programs need only two machine-related constants: an epsilon, whose role was described earlier, and the REAL value of the largest valid integer. We have isolated these as the variables EPSI and MAXINT in the COMMON block NUMBRS so that they can be set once at initialization. The initialization subroutine, CINIT, takes care of this.

CINIT, which should be called before any of the other FORTRAN routines in this book, also sets several other variables that may vary from installation to installation or from run to run:

OUNIT   the FORTRAN unit number for output (often unit 6),
PMIN   the left margin in the output line,
PMAX   the right margin in the output line.

In CINIT, the corresponding subroutine arguments all begin with the letter I to indicate that they are initialization values. CINIT performs several basic checks on these and then completes the initialization process. In the course of a

```
       SUBROUTINE CINIT(IOUNIT, IPMIN, IPMAX, IEPSI, IMAXIN, ERR)
C
       INTEGER IOUNIT, IPMIN, IPMAX, IMAXIN, EPR
       REAL IEPSI
C
C   INITIALIZATION, TO BE CALLED AT START OF ANY  MAIN PROGRAM
C   WHICH CALLS ONE OF THE EDA SUBROUTINES (EITHER DIRECTLY OR
C   INDIRECTLY).
C
C   IOUNIT   IS THE NUMBER OF THE UNIT TO WHICH OUTPUT IS DIRECTED.
C   IPMIN    IS THE LEFT MARGIN.
C   IPMAX    IS THE RIGHT MARGIN.
C   IEPSI    IS THE MACHINE-RELATED EPSILON.
C   IMAXIN   IS THE MAXIMUM PERMITTED INTEGER VALUE
C
C   ERR IS THE (USUAL) ERROR FLAG, TO INDICATE WHETHER
C   THE ROUTINE EXECUTED SUCCESSFULLY.
C
       COMMON /CHRBUF/ P , PMAX, PMIN, OUTPTR, MAXPTR, OUNIT
       COMMON /NUMBRS/ EPSI, MAXINT
C
       INTEGER P(130), PMAX, PMIN, OUTPTR, MAXPTR, OUNIT
       REAL EPSI, MAXINT
C
C   LOCAL VARIABLES
C
       INTEGER BLANK, I
       DATA BLANK /1H /
C
C
       ERR = 6
       IF(IPMIN .LT. 1) GO TO 999
       IF(IPMAX .GT. 130) GO TO 999
       IF(IPMAX .LE. IPMIN) GO TO 999
       ERR = 7
       IF((1.0 + IEPSI) .LE. 1.0) GO TO 999
       ERR = 0
       OUNIT = IOUNIT
       PMIN = IPMIN
       OUTPTR = IPMIN
       MAXPTR = IPMIN
       PMAX = IPMAX
       EPSI = IEPSI
       MAXINT = FLOAT(IMAXIN)
C
       DO 50 I = 1, 130
         P(I) = BLANK
   50 CONTINUE
C
  999 RETURN
       END
```

sequence of analyses, using several of the programs in this book, a user may reset the initialization variables by again calling CINIT. Of course, this causes the previous values of these variables to be lost, and it causes the output line to be set to all blanks, but it has no other side effects.

Stream Output

FORTRAN requires that the programmer specify the contents and format of a line of output, essentially when the program is written. (While it is possible for a running program to read a format specification or to construct one, it is extremely difficult to program this in a portable way.) Because EDA displays, such as the boxplot, depend heavily on the data, we usually can be no more specific about the output format than to say that a line will contain a number of characters—some digits, some symbols, and some blank spaces. As the program executes, it must determine the format for a line and the character that occupies each position on the line. For example, stem-and-leaf displays come in three different formats, and each requires different characters in special positions on the line. Thus the program needs to build each output line a few characters at a time.

This style of output—allowing the program to determine the format and contents of the output line as it goes along—is known as *stream output*. Because such output capabilities are not a part of the FORTRAN language, we have written special subroutines to simulate (in a rudimentary but portable way) the features that we need to produce our EDA displays. Often, we have used standard FORTRAN output.

The important variables for our stream output subroutines reside in the COMMON block CHRBUF. At the heart of our simple stream output is the array P, in which we construct a line of output. Our initialization routine, CINIT, sets P to all blanks. Any routine needing to construct an output line can do so by storing characters (alphabetic, numeric, or special symbols) in P; this is usually done with the subroutines PUTCHR and PUTNUM. When the line is complete, the routine PRINT writes out the contents of P and resets P to blanks.

The routine PUTCHR places a character in P, either at the position specified by the argument POSN or at the next available position (if POSN is zero). PUTCHR keeps track of the last print position used and the rightmost non-blank position in the line.

The routine PUTNUM places into P the characters for an integer, N. The calling program must specify the width, W, of the field (number of characters) where the number should appear, and its starting position on the line. PUTNUM translates the integer into the appropriate sequence of numerals and uses PUTCHR to place them in P. Applications of PUTNUM include placing the depth counts and the stems on each line of a stem-and-leaf display.

Finally, the integer function WDTHOF receives an integer, I, and returns the number of characters (including a minus sign if I is negative) required to print it. We use this information in printing the depth counts and stems in a stem-and-leaf display.

Conventions

To promote clarity of these programs and to preserve their portability, we have followed several conventions. None of these has especially sweeping consequences, but we list them here so that they will be clear to the reader and user.

Input/Output.   Our subroutines do no input. Reading of data is the responsibility of the user, who is in the best position to deal with features of the input process that may depend on the particular version of FORTRAN or on the devices where data are stored. It is customary to isolate output operations so that they do not appear in computational subroutines. We have done this where appropriate; but, of course, it makes no sense when the EDA technique is primarily a display (as in stem-and-leaf, boxplot, condensed plotting, and coded tables).

Scratch Storage.   When a technique uses temporary storage whose size depends on the number of data values, our routines are structured so that the user supplies this storage through the argument list. (PLOT, for example, requires two work arrays of length N because it must sort the data points into order on *y* while preserving the (*x,y*) pairs.) In this way we avoid any built-in restriction on the amount of data that can be handled, and we make it straightforward to accommodate the storage limitations that the user's system may impose.

Characters.   When we must work with characters, we store them, one character to the word, in INTEGER variables or arrays. This may waste a certain amount of space, but it is strongly preferable to dealing with heavy dependence on the number of characters that can be stored in a word on the user's particular machine. It further avoids the arithmetic that would be required to pack and unpack characters stored several to the word. The character set that we have used is the bare minimum FORTRAN character set: the 26 letters, the 10 digits, the 9 symbols = + − * / ( ) , . and the blank space. This facilitates portability, but it is not much to work with in building displays. In BASIC we are able to assume the much larger ASCII character set, and the advantages are evident when one compares the BASIC and FORTRAN versions of the displays.

Dimensioning in Subroutines. When a subroutine argument is an array, our declaration for it uses its actual dimensions, as in "REAL Y(N), . . ." in STMNLF. We have not used "dummy" dimensions, as in "REAL A(1)" seen in some programs.

Errors. We attempt to detect a variety of errors that a user might make, and we communicate information on them through the INTEGER variable ERR, which appears as the last argument of many of the subroutines. If no error condition exists, ERR has the value 0. Otherwise, a positive value identifies the error condition. (These error numbers are defined in Exhibit C–1.)

Exhibit C–1 FORTRAN Program Error Codes

Code	Subroutine	Meaning
1	SORT	$N \leq 0$; nothing to sort
2	PSORT	$N \leq 0$; nothing to sort
3	INTFN	$X >$ MAXINT; argument passed is too large to be "fixed" as an integer variable
4	PUTCHR	Illegal character code
5	PUTNUM	Number won't fit in space provided
6	CINIT	Violated $0 <$ IPMIN $<$ IPMAX < 130 in setting page margins
7	CINIT	EPSI too small; $1.0 +$ EPSI $= 1.0$
8	NPOSW	No room allowed for plot
9	NPOSW	HI $<$ LOW
11	STMNLF	$N \leq 1$
12	STEMP	Bad internal value—bad nice numbers?
13	STMNLF	Page too narrow for display
21	LVALS	Violated $2 \leq N \leq 24576$
22	LVPRNT	Violated $3 \leq$ NLV ≤ 15; too many letter values
23	LVPRNT	Page width < 64 positions, not enough room

Exhibit C–1   (continued)

Code	Subroutine	Meaning
31	BOXES	$N \leq 1$
41	PLOT	$N < 5$
42	PLOT	Violated $5 \leq$ LINES ≤ 40 or $1 \leq$ CHRS ≤ 10
44	PLOT	XMIN > XMAX
45	PLOT	YMIN > YMAX —Errors 44 and 45 are possible if incorrect plot bounds have been specified in the subroutine call.
51	RLINE	$N < 6$
52	RLINE	No iterations specified
53	RLINE	All *x*-values equal; no line possible
54	RLINE	Split is too uneven for resistance
61	RSM	$N < 7$
62	RUNMED	Insufficient workspace room
63	RUNMED	Internal error—error in sort program? This error can occur if a system sort utility is substituted for the supplied SORT subroutine, but used incorrectly.
71	CTBL	Zero dimensions for table
72	CTBL	Too many columns to fit on page
81	MEDPOL or TWCVS	Zero dimensions for table
82	MEDPOL	No half-steps specified
83	MEDPOL	Illegal start parameters
85	MEDPOL	Table is empty
88	TWCVS	Zero grand effect; can't compute comparison values
91	RGCOMP	$L \leq 2$; too few bins
92	RGCOMP	One of the hinges falls in the left-open bin or in the right-open bin
93	RGPRNT	Page too narrow for rootogram table
94	RGPRNT	Room for rootogram table but not for graphic display

Exits. Each of our subroutines has a single exit, the RETURN statement immediately preceding the END statement. In most subroutines this RETURN bears the statement number 999.

Output FORMAT statements. We place each FORMAT statement immediately after the first WRITE statement that uses it. For our programs, which do not use the same FORMAT statement in many different and widely separated WRITE statements and often rely on the stream output routines described earlier, this leads to much better readability than if we grouped all FORMAT statements at the end of the subroutine.

Declared Identifiers. We do not rely on "implicit typing" to determine (according to its first letter) whether an identifier is INTEGER or REAL. Instead, we explicitly declare all the identifiers used in each subprogram, except for the standard FORTRAN functions. We strongly endorse this practice, which a few FORTRAN compilers support by issuing a warning message for any undeclared identifier, because it aids greatly in eliminating misspelled names. (The PFORT Verifier, for example, lists all the identifiers in each program unit, so that such errors stand out.)

Indentation. We find that it is generally easier to follow the logic of a program when statements within a DO loop or following an IF statement are indented slightly, and we have used this device throughout our programs.

Reference

Isaacs, Gerald L. 1976. "BASIC REVISITED, An Update to Interdialect Translatability of the BASIC Programming Language." CONDUIT, The University of Iowa, Iowa City.

Ryder, B.G. 1974. "The PFORT Verifier." *Software—Practice and Experience* 4:359–377.

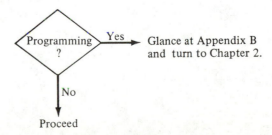

Appendix D
Minitab Implementation

The FORTRAN programs presented in this book have been incorporated into the Minitab statistics package. This appendix gives the syntax of the Minitab commands for exploratory data analysis techniques. It assumes a familiarity with the Minitab package. Readers unfamiliar with Minitab should read the *Minitab Student Handbook,* (Ryan, Joiner, and Ryan, 1976) or the *Minitab Reference Manual* (Ryan, Joiner, and Ryan, 1981).

The commands given here may change slightly as the Minitab system changes. For details of the current status of the system, use the Minitab HELP command or refer to the latest edition of the *Minitab Reference Manual*.

Minitab is an excellent environment for exploratory data analysis computing, especially when used interactively. Minitab works with data kept in a computer worksheet, where the data values are stored in columns designated C1, C2, . . . , or in matrices designated M1, M2, ... Single numbers can be stored in constants designated K1, K2, ... Although variables in the worksheet may have names (which are surrounded by quotes, like 'INCOME' or 'RACE'), the command syntax usually shows the generic names **C** for column, **K** for constant, and **M** for matrix. Thus, the command specified as

STEM C

indicates a command in which **C** is to be replaced by any column identifier (for example, C3, C17, 'MONEY').

When portions of a Minitab command line are optional, we enclose these portions in square brackets. Some commands allow subcommands that modify the main command. When a subcommand follows the main command line, Minitab requires that the main command line end with a semicolon. Each subsequent subcommand line ends with a semicolon, up to the final subcommand, which ends with a period.

Minitab command lines may contain free text, which further describes the operation performed but has no effect on Minitab. The command descriptions in this appendix take advantage of this feature to include brief explanations of the commands and subcommands. Only the portions of the command descriptions in **boldface** are actually required.

References

Ryan, Thomas A., Brian L. Joiner, and Barbara F. Ryan. 1976. *Minitab Student Handbook*. Boston: Duxbury Press.

Ryan, Thomas A., Brian L. Joiner, and Barbara F. Ryan. 1981. *Minitab Reference Manual*. University Park, Pennsylvania: Minitab Project, The Pennsylvania State University.

D.1 Stem-and-Leaf Displays

STEM-AND-LEAF DISPLAY OF C C
>   Gives a separate stem-and-leaf display for each column named.

Optional Subcommands

TRIM OUTLIERS (default)
>   Scale to the adjacent values.

NOTRIM
>   Scale to the extremes of the data—no HI or LO stems.

Examples
>   STEM 'RAINPH'
>
>   STEM 'HC' 'JANTMP'
>
>   STEM 'HC';
>   NOTRIM.

D.2 Letter-Value Displays

LVALS OF C [PUT LETTER VALUES IN C [MIDS IN C [SPREADS IN C]]]
>   This command prints a letter-value display. Optionally, the letter values, mids, and spreads can be stored in specified columns. The column of letter values will be roughly twice as long as the columns of mids and spreads, and will start with the low extreme and proceed in order to the high extreme.

Examples
>   LVALS OF 'NJCOUNT'
>
>   LVALS OF 'MSPRAIN' PUB IN C1, 'MIDS', 'SPREADS'

D.3   Boxplots

BOXPLOTS FOR **C** [LEVELS IN **C**]

The levels column is the same length as the data column. It labels each data value with an integer that identifies the level, subscript, group, or cell to which the value belongs. A boxplot will be produced for the data in each level, all on the same scale. If no levels column is specified, a single boxplot is produced.

Levels.   The levels must be integers between -1000 and 1000. Up to 100 distinct levels are allowed.

Optional Subcommands

The following subcommands control the plots.

LINES = **K**

K is the number of lines used to print a box. K can be 1 or 3. If LINES is not specified, K is assumed to be 3.

NOTCH THE BOXPLOTS TO INDICATE CONFIDENCE INTERVALS FOR THE MEDIAN

NONOTCH (default)

LEVELS K, . . . , K [FOR **C**]

This specifies what subscript levels (cells, group numbers) are to be used, and in what order. This subcommand can be used (a) to arrange the groups in a certain order, (b) to get boxplots for only some groups, or (c) to include (empty) boxplots for groups which are theoretically possible but are not present in the sample.

Example

BOXPLOTS FOR 'IRSAUDIT', LEVELS IN 'REGION';
    NOTCH.

D.4   Condensed Plotting

CPLOT Y IN C VS X IN C

This command produces a condensed plot.

Optional Subcommands

LINES = **K**

Specifies how many lines (up to 40) the plot should take. (Default is 10.)

CHARACTERS = **K**

Specifies how many codes should be used, and thus how many subdivisions each line is to be cut into. K can be between 1 and 10. (Default is 10.)

XBOUNDS K TO **K**

Specifies the range in the x direction of the data to be plotted. Data values beyond the specified range will appear as outliers in the plot.

YBOUNDS K TO **K**

Specifies the range in the y direction of the data to be plotted.

Plot Width

The width of the plot can be changed by using the Minitab OUTPUT-WIDTH command prior to the CPLOT command.

Examples

```
CPLOT 'BIRTHS' BY 'YEAR';
    LINES = 40;
    CHARACTERS = 1.

CPLOT 'BIRTHS' BY 'YEAR';
    LINES = 10;
    YBOUNDS 1940 TO 1960.
```

D.5   Resistant Lines

RLINE Y IN **C**, X IN **C** [PUT RESIDS INTO **C** [PRED INTO **C** [COEFF INTO **C**]]]
Fits a resistant line to the data.

Optional Subcommands

MAXITER = **K**
Specifies the maximum number of iterations. (Default is 10.)

HALFSLOPES STORED, LEFT HALFSLOPE IN **K**, RIGHT HALFSLOPE IN **K**

REPORT EACH ITERATION (default)

NOREPORT
Minitab will print only the final solution; it will not report each iteration.

Missing Data

If either x or y is equal to the missing value code, *, for an observation, the observation is not used in fitting the line. If x is missing, the predicted value and residual are set to *. If x is not missing and y is missing, the predicted value is computed as usual, and the residual is set to *. *Note:* At least 6 (non-missing) data points are needed.

Examples
RLINE 'CANCR' VS 'TEMP' RESIDS IN 'RESID';
    MAXITER = 20.

RLINE 'MPG' ON 'DISP' RESIDS IN C1, PRED IN C2;
    HALFSLOPES K1 K2.

D.6 Resistant Smoothing

RSMOOTH O, PUT ROUGH IN **O**, SMOOTH IN **C**

Applies a resistant smoother to sequence data. The rows are assumed to be in sequence order. (Note that the order in which the storage columns are specified corresponds to the residuals and predicted values in regression, resistant line, median polish, and so on.)

Note: This command produces no output. The smooth and rough may be plotted with the Minitab TSPLOT command.

Optional Subcommands

SMOOTH 3RSSH, TWICE (specifies this smoother)

SMOOTH 4253H, TWICE (default)

Missing Data

Missing observations are allowed at the beginning and end of the series only. That is, missing values cannot come between valid data values. The results (both smooth and rough) for rows corresponding to missing data are set to the missing value.

Examples

RSMOOTH 'COWTMP' PUT ROUGH IN 'ROU', SMOOTH IN 'SMO'

RSMOOTH 'COWTMP';
    SMOOTH 3RSSH.

D.7 Coded Tables

CTABLE OF DATA IN **C**, ROW LEVELS IN **C**, COLUMN LEVELS IN **C**

Prints a coded table of the data. The levels columns specify rows and columns of the table.

Levels. Levels must be integers between -1000 and 1000. Each levels column can contain up to 100 distinct values.

Optional Subcommands

LEVELS K, ..., **K** FOR **C**

This subcommand allows reordering of the specified column of row or column levels. The table will be printed with the specified levels in the prescribed order. Note that a level value that does not appear in the specified column of levels may be specified in a LEVELS subcommand. It will cause an empty row or column to appear in the table. Two LEVELS subcommands may be used, one to specify an order for rows, and one to specify an order for columns.

MAXIMUM OF MULTIPLE VALUES IN A CELL SHOULD BE CODED

MINIMUM OF MULTIPLE VALUES IN A CELL SHOULD BE CODED

EXTREME OF MULTIPLE VALUES IN A CELL SHOULD BE CODED

These three subcommands may be used when two or more data values have the same row and column numbers—that is, when a cell of the table contains more than one data value. The subcommands specify what feature of the cell is to be coded. The default is EXTREME.

Examples

CTABLE OF 'MORT', LEVELS IN 'CAUSE', 'SMOKE'

CTABLE OF 'SURVTIME', LEVELS IN 'POISON', 'TREAT';
    LEVELS 2, 3, 1 IN 'TREAT';
    MAXIMUM.

D.8   Median Polish

MPOLISH C, LEVELS IN C, C [RESIDS INTO C [PRED INTO C]]

Uses median polish to fit an additive model to a two-way table.

Levels.   Levels must be integers between -1000 and 1000. Each levels column can contain up to 100 distinct values.

Optional Subcommands

ROWS FIRST (default)

Begin by finding and subtracting row medians.

COLUMNS FIRST

ITERATIONS = K

Number of half-steps to be performed. (Default is 4.)

COMPARISON VALUES INTO C

EFFECTS STORED, COMMON IN K, ROW EFFECTS IN C, COLUMN EFFECTS IN C

LEVELS K, . . . , K FOR C

This subcommand reorders the levels or specifies which rows or columns of the table are to be analyzed. Its use is similar to the LEVELS subcommand of CTABLE or BOXPLOT.

Output

The **MPOLISH** command prints a table of residuals bordered on the right by row effects and on the bottom by column effects, with the common term at the lower right. In addition, the fitted values can be printed using the TABLE command in Minitab. The residuals might be displayed in a coded table by using the CTABLE command, or they might be plotted against the comparison values and fitted with a resistant line.

Example

```
MPOLISH 'DEATHS' BY 'SMOKE' AND 'CAUSE', RESIDS IN C1, PRED IN C2;
    ITERATIONS  = 6;
    COMPARISON VALUES IN 'COMP';
    EFFECTS IN K9. 'REFF'. 'CEFF'.
```

D.9   Suspended Rootograms

ROOTOGRAM [FOR DATA IN **C** [USING BIN BOUNDARIES IN **C**]]

Prints a suspended rootogram for the data. If no bin boundaries are specified, the program determines them by a method similar to the scaling algorithm of the stem-and-leaf display. If bin boundaries are specified, the program computes bin counts by counting the number of data values less than the smallest bin boundary, between the first and second boundaries, . . . , greater than the largest bin boundary. Each bin but the last contains numbers less than or equal to its upper boundary.

Optional Subcommands to Store Results

BOUNDARIES STORED IN **C**

If bin boundaries have been determined automatically, this subcommand stores them in the specified column.

DRRS STORED IN **C**

Stores the double-root residuals.

FITTED VALUES STORED IN **C**

Stores the fitted bin counts (which need not be integers) in the specified column.

COUNTS STORED IN **C**

Stores the observed bin counts in the specified column.

Optional Subcommand to Use Bin Frequencies

FREQUENCIES IN **C** [FOR BINS WHOSE BOUNDARIES ARE IN **C**]

This subcommand specifies a data column of bin frequency counts and the corresponding bin boundaries. It should be used when the data are available as frequencies recorded bin by bin. (This subcommand does not use columns specified in the main command line. Minitab will warn of an error if the FREQUENCIES subcommand is used when columns are specified in the main command line.) The first bin count is assumed to be for the half-open bin below the lowest bin boundary, and must be zero if no data values fall below the lowest bin boundary. The last count corresponds to the half-open bin about the highest bin boundary. The last count must be zero if no data values fall above the highest bin boundary. Thus the column of bin frequencies has one more entry than does the column of bin boundaries. If no bin boundaries are specified, the frequencies are assumed to be for bins of equal width, and the bin width is arbitrarily taken to be 1.

Optional Subcommands to Control the Fitted Shape

MEAN = K

This subcommand overrides the automatic estimation of the mean of the data and uses the specified mean in fitting the Gaussian comparison curve.

STDEV = K

This subcommand overrides the automatic estimation of the standard deviation of the data and uses the specified standard deviation in fitting the Gaussian comparison curve.

These two subcommands can be used together to specify a particular Gaussian distribution for calculating the fitted counts. This may be useful if there are theoretical or other reasons for wishing to compare the data to that particular Gaussian distribution.

Note: The rootogram output will be affected by the OUTPUT-WIDTH command in Minitab. If the available output width is less than 65 spaces, the observed and fitted values and the double-root residuals will be printed, but the rootogram will not be displayed.

Example
```
ROOTOGRAM;
    FREQUENCIES IN 'SOLDRS' BY 'CHEST'.
```

Index

Special symbols come first, in an order similar to the order established by the ASCII character set. These are followed by the numeric symbols associated with resistant smoothing.

Page numbers in **boldface** indicate the definition of a term or concept or the full tabulation of a data set.

347